AQA(A) A2

UNIT
3

Psychology

Topics in Psychology (2):
Biological Rhythms and Sleep, Relationships,
Aggression, Cognition and Development

Jean-Marc Lawton

Philip Allan Updates, an imprint of Hodder Education, an Hachette UK company, Market Place, Deddington, Oxfordshire OX15 0SE

Orders

Bookpoint Ltd, 130 Milton Park, Abingdon, Oxfordshire OX14 4SB
tel: 01235 827720
fax: 01235 400454
e-mail: uk.orders@bookpoint.co.uk

Lines are open 9.00 a.m.–5.00 p.m., Monday to Saturday, with a 24-hour message answering service. You can also order through the Philip Allan Updates website: www.philipallan.co.uk

© Philip Allan Updates 2009

ISBN 978-0-340-99175-6

First printed 2009
Impression number 5 4 3 2 1
Year 2014 2013 2012 2011 2010 2009

This guide has been written specifically to support students preparing for the AQA Specification A A2 Psychology Unit 3 examination. The content has been neither approved nor endorsed by AQA and remains the sole responsibility of the author.

Typeset by Phoenix Photosetting, Chatham, Kent
Printed by MPG Books, Bodmin

Hachette UK's policy is to use papers that are natural, renewable and recyclable products and made from wood grown in sustainable forests. The logging and manufacturing processes are expected to conform to the environmental regulations of the country of origin.

7441 3

A2 Psychology

Contents

Introduction

■ ■ ■

Content Guidance

■ ■ ■

Questions and Answers

Introduction

About this guide

This guide will help you prepare for the AQA(A) A2 Psychology Unit 3 examination. It covers four of the eight topics listed in the specification; the other four topics are covered in the preceding book, AQA(A) A2 Psychology, Unit 3: Topics in Psychology (1), ISBN 978-0-340-99176-3. This publication is not intended as a revision guide or textbook, but rather as a support device to revision and learning. Therefore, this guide looks first at the specification content and how it is examined, and second at how answers of varying quality are assessed.

- The specification content for each topic is fully explained so that you will understand what is required from you in the examination (though other content may be equally appropriate).
- Content appropriate to each topic is outlined, to an extent that it is possible to construct an answer to possible questions set on that topic.
- A sample question for each topic is provided, along with an explanation of its requirements.
- Sample answers are also provided, along with examiner comments explaining the strengths and limitations of each answer.

Using this guide

You can use this guide in a variety of ways.

- **During your course**, each time you start a new topic (e.g. biological rhythms and sleep), use the unit guide to give you a quick overview of what is involved. Re-read each topic at regular intervals as you are studying it in class.
- **When you start revising**, use the unit guide to review the specification areas you have studied (e.g. disorders of sleep). Use the unit guide to refresh your learning and consolidate your knowledge of each of the four Unit 3 topics covered.
- **When practising for the exam**, use the Questions and Answers section. Ideally you should attempt the questions yourself before reading the sample answers and examiner's comments, and then compare your answer with the one given. If you don't have time for this, you should at least make brief plans that you could use as the basis of an answer to each question. Study the sample answer and the examiner's comments and then add the key points from them to your own answer or plan.

The examination

The Unit 3 exam lasts 1 hour and 30 minutes and there will be eight essay-style questions (one drawn from each of the eight topics), from which you must select and answer three. No section is compulsory, so you will have a completely free choice.

Each question is worth 25 marks overall, although some questions may be split into parts. You are guaranteed one question on each of the eight topics that comprise introduction

Unit 3, but to ensure that you can answer the question that has been set on a topic you have covered, you must have studied and revised all the subject content listed in the specification for that topic.

This paper will account for 50% of the total A2 marks and 25% of the total A-level (AS + A2).

Assessment objectives

In this psychology examination, three sets of skills or 'assessment objectives' are tested: AO1, AO2 and AO3.

AO1 (assessment objective 1) concerns questions designed to test your knowledge and understanding of psychological theories, terminology, concepts, studies and methods. You should be able to:
- recognise, recall and show understanding of knowledge
- select, organise and communicate relevant information in a variety of forms
- present and organise material clearly
- use relevant psychological terminology

AO2 (assessment objective 2) concerns questions designed to test your knowledge and understanding of the application of knowledge via analysis and evaluation of psychological theories, concepts, studies and methods. You should be able to:
- analyse and evaluate knowledge and processes
- apply knowledge and processes to novel situations, including those relating to issues
- assess the validity, reliability and credibility of information

AO3 (assessment objective 3) concerns questions designed to test your knowledge and application of knowledge and understanding of how psychology as a science works. You should be able to:
- describe ethical, safe and skilful practical techniques and processes and the appropriate selection of qualitative and quantitative methods
- know how to make, record and communicate reliable and valid observations and measurements with appropriate accuracy and precision, through using primary and secondary sources
- analyse, interpret, explain and evaluate the methodology, results and impact of experimental and investigative activities in a variety of ways

For each question there will be 9 AO1 marks, 12 AO2 marks and 4 AO3 marks (a grand total of 75 possible marks on offer). You may sit this exam either in January or June of each year.

Good practice

For AO1 you should:
- avoid 'storytelling' or 'commonsense' answers that lack psychological content
- give some depth to your answers and not just provide a list of points
- achieve a balance between the breadth and depth of your answer
- make your answer coherent; therefore it should be clearly written and have continuity, so that it does not read as a series of unconnected comments

For AO2/AO3 you should:

- elaborate upon evaluative points in order to construct an effective commentary
- where possible and appropriate, make use of both negative and positive criticism — for example, methodological faults and practical applications
- draw conclusions and interpretations from your AO1 material
- select material carefully so that it is specifically directed at the question, rather than just forming a general answer on the topic area
- avoid overuse of generic evaluation, such as the repetitive detailing of methodological strengths and weaknesses of all research studies included in your answer
- present arguments and evaluations with clarity
- ensure that you have included within your evaluation and analysis material on relevant issues/debates/approaches — for example, ethical issues and/or the nature–nurture debate

Mark band descriptors

AO1 mark bands (5 marks)

5–4 marks	Outline is reasonably thorough, coherent and accurate.
3–2 marks	Outline is limited, reasonably coherent and generally accurate.
1 mark	Outline is weak and muddled.
0 marks	No creditworthy material is apparent.

AO2 mark bands (9 marks)

9–8 marks	**Sound** Knowledge and understanding are accurate and well detailed. A good range of relevant material has been presented. There is substantial evidence of breadth/depth. Organisation and structure of the answer are coherent.
7–5 marks	**Reasonable** Knowledge and understanding are generally accurate and reasonably detailed. A range of relevant material has been presented. There is evidence of breadth and/or depth. Organisation and structure of the answer are reasonably coherent.
4–3 marks	**Basic** Knowledge and understanding are basic/relatively superficial. A restricted range of material has been presented. Organisation and structure of the answer are basic.
2–1 marks	**Rudimentary** Knowledge and understanding are rudimentary and may be muddled and/or inaccurate. The material presented may be brief or largely irrelevant. The answer lacks organisation and structure.
0 marks	No creditworthy material is apparent.

AO3 mark bands (16 marks)

16–13 marks	**Effective** Evaluation shows sound analysis and understanding. Answer is well focused and displays coherent elaboration and/or a clear line of argument is apparent. Effective use of issues/debates/approaches. There is substantial evidence of synopticity. Well-structured ideas are expressed clearly and fluently. There is consistent effective use of psychological terminology and appropriate use of grammar, spelling and punctuation.
12–9 marks	**Reasonable** Evaluation shows reasonable analysis and evaluation. A generally focused answer that displays reasonable elaboration and/or line of argument is apparent. A reasonably effective use of issues/debates/approaches. There is evidence of synopticity. Most ideas are appropriately structured and expressed clearly. There is appropriate use of psychological terminology and there are some minor errors of grammar; spelling and punctuation only occasionally compromise meaning.
8–5 marks	**Basic** Evaluation and analysis show basic, superficial understanding. An answer that is sometimes focused and has some evidence of elaboration. There is a superficial use of issues/debates/approaches. There is some evidence of synopticity. The expression of ideas lacks clarity. There is limited use of psychological terminology and errors of grammar, spelling and punctuation are intrusive.
4–1 marks	**Rudimentary** Evaluation and analysis are rudimentary, showing very limited understanding. A weak, muddled and incomplete answer. Material is not used effectively and may be mainly irrelevant. Any reference to issues/debates/approaches is muddled or inaccurate. There is little or no evidence of synopticity. The expression of ideas is deficient, demonstrating confusion and ambiguity. The answer lacks structure and may be just a series of unconnected points. There are errors in grammar, spelling and punctuation that are frequent and intrusive.
0 marks	No creditworthy material is evident.

Explanation of examination injunctions

AO1

Outline: provide brief details without explanation

Describe: provide a detailed account without explanation

AO2

Evaluate: assess the value/effectiveness

Discuss: provide a reasoned, balanced account

Unit 3 topics

The Content Guidance section covers the main issues, themes and debates that you need to be familiar with in the study of the four A2 topics covered. The aim is to give an overview of these topics and to outline the key points you need to know in order to tackle the unit examination. However, it should be remembered that, for a full and adequate knowledge of the subject matter of this unit, you also need to study your textbook(s) and the notes you make during your course. You can use the points made in this section both to organise your own notes and studies, and as a revision aid when preparing for the examination.

In the four topics covered in this guide, the AQA specification requires you to study the following major areas.

- **Biological rhythms and sleep**
  - Biological rhythms
  - Sleep states
  - Disorders of sleep
- **Relationships**
  - Formation, maintenance and breakdown of romantic relationships
  - Human reproductive behaviour
  - Effects of early experience and culture on adult relationships
- **Aggression**
  - Social psychological approaches to explaining aggression
  - Biological explanations of aggression
  - Aggression as an adaptive response
- **Cognition and development**
  - Development of thinking
  - Development of moral understanding
  - Development of social cognition

Content
Guidance

In this section, guidance is given on each of the three subsections of the four topics covered by this unit guide.

Each subsection starts by providing an outline and explanation of what the specification demands. This is then followed by a more detailed examination of the theories, research studies and evaluative points of which each subsection is made up.

Wherever appropriate, a general pattern for each topic will be followed, providing an outline and explanation of what the specification demands. The subject matter will then be described, research evidence given and finally further evaluative points made.

It is important to remember that research evidence can be used either as descriptive material when answering examination questions (AO1) or as evaluation (AO2/AO3). It is advisable to learn how to make use of such material in an evaluative way — for instance, by using such wording as 'this supports', 'this suggests' and so on.

For the research quoted, names of researchers and publication dates are given. Full references for these should be available in textbooks and via the internet if you wish to study them further.

Biological rhythms and sleep

Biological rhythms

Specification

- *Circadian, infradian and ultradian rhythms, including the role of endogenous pacemakers and exogenous zeitgebers*
- *Consequences of disrupting biological rhythms, for example shift-work, jet lag*

Circadian, infradian and ultradian biological rhythms are specifically identified, meaning there is a requirement for them to be studied, as they could form an explicit examination question. This includes both endogenous pacemakers and exogenous zeitgebers, as both are directly named.

The second requirement is to have knowledge of the consequences of disrupting biological rhythms, so that they could be both described and evaluated. Shift-work and jet lag are given as examples, so they would not be named explicitly in an examination question, and any other relevant material would be equally acceptable.

Biological rhythms

Biological rhythms are cyclical behaviours, i.e. repeated periodically, controlled either by **endogenous pacemakers** (internal biological clocks regulating biological functioning), or by **exogenous zeitgebers** (external/environmental cues, such as seasonal changes).

Circadian rhythms

These are biological cycles lasting around 24 hours, like the **human sleep–wake cycle**, which is usually facilitated by time-checks and regular events, such as meal times. There is a free-running cycle controlled by an endogenous pacemaker working as a 'body clock'. Another circadian rhythm is **body temperature**, rising and declining as an indicator of metabolic rate, 4.00 a.m. being the lowest point.

Research
- Siffre (1972) spent 6 months in a cave with no time cues. Artificial lights came on when he was awake. He settled into a sleep–wake cycle of 25–30 hours. After 179 days he thought 151 days had passed, implying that endogenous pacemakers exert a strong influence on circadian rhythms, although the use of artificial light may have been a confounding variable.
- Aschoff and Weber (1965) placed participants in a bunker with no natural light. They settled into a sleep–wake cycle of between 25 and 27 hours, giving support to the Siffre study, again suggesting that endogenous pacemakers control the sleep–wake cycle in the absence of light cues and that light seems necessary to coordinate the biological clock with the external environment.

- Folkard et al. (1985) isolated 12 participants from natural light for 3 weeks, manipulating the clock so that only 22 hours passed a day. Eleven participants kept pace with the clock, showing the strength of the circadian rhythm as a free-running cycle.

Evaluation

- The research suggests practical applications, such as designing timetables around the optimal times to study/work, or when best to take medicines.
- Individual differences exist in sleep–wake cycles. Duffy et al. (2000) found that early risers prefer 6 a.m. to 10 p.m. and later risers prefer 10 a.m. to 1 a.m. Aschoff and Weber (1976) found in isolation studies that some participants maintain normal cycles, while others strongly differ.
- Isolation studies have few participants, making generalisation problematic.
- Research suggests that endogenous pacemakers do exist and are regulated to some extent by exogenous zeitgebers.

Infradian rhythms

These are biological cycles lasting more than 24 hours, such as the menstrual cycle, which is regulated by hormone secretions. These were originally thought to be controlled by the hypothalamus acting as an endogenous pacemaker, but evidence now shows that exogenous zeitgebers play a part too. Infradian rhythms include **circannual rhythms** occurring once a year, such as hibernation.

Research

- McClintock and Stern (1998) found that women who inhaled fumes from armpit pads of other women who were about to ovulate had shorter menstrual cycles, while inhaling fumes from women who had just ovulated made menstrual cycles longer. It seems that pheromones in the donor's sweat affect the recipient's infradian rhythm, suggesting that exogenous zeitgebers have a regulating affect.
- Russell et al. (1980) applied underarm sweat of donor women to upper lips of female participants. They found that menstrual cycles became synchronised, supporting McClintock and Stern's findings and suggesting that pheromones act as exogenous zeitgebers.
- Reinberg (1967) reported on a woman who spent 3 months in a cave without natural lighting. Her menstrual cycle shortened to 25.7 days, implying that infradian rhythms can be influenced by exogenous zeitgebers such as light.
- Rosenzweig et al. (1999) reported on Seasonal Affective Disorder (SAD), where for some people winter darkness brings low moods. It has been associated with darkness stimulating the production of melatonin, a hormone associated with the regulation of sleep, stressing the importance of light as an exogenous zeitgeber.

Evaluation

- The effects of pheromones on women's menstrual cycles may explain why women living together, such as nuns and nurses, tend to have synchronised periods. Turke (1984) believes it has an evolutionary significance, allowing women living together to synchronise pregnancies and share childcaring duties. Women working in close proximity to men often have shorter cycles, possibly as a response to male pheromones bestowing an evolutionary advantage in giving more opportunities to get pregnant.

- The results found by McClintock concerning synchronisation of menstrual periods are explainable as random occurrences and do not form a significant difference statistically. Moreover, women's cycles are not universal, which may invalidate findings. What is needed is evidence that women with different cycle lengths show synchronisation.
- Research into SAD has led to the development of successful light therapies to treat the condition.

Ultradian rhythms

These are biological cycles lasting less than 24 hours, such as the cycle of brain activity during sleep. Sleep has several stages occurring through the night, lasting for about 1 hour in infancy and 90 minutes by adolescence.

Research

- Rechtschaffen and Kales (1968) measured electrical activity of the brain with an electroencephalogram (EEG), finding different patterns of activity at different times of sleep.
- Klein and Armitage (1979) tested participants on verbal and spatial tasks, finding that performance was related to a 96-minute cycle, very similar to the sleep cycle.
- Gerkema and Dann (1985) found that ultradian rhythms tend to be correlated with brain and body size, with larger animals having longer cycles.

Evaluation

- Creating lesions to brain areas that control circadian rhythms has no effect on behaviours with an ultradian rhythm, suggesting that circadian and ultradian rhythms have different controlling mechanisms.
- A lot of the research into ultradian rhythms has involved animals, creating problems with generalising findings to humans.

Role of endogenous pacemakers

The main pacemaker is the superchiasmatic nucleus (SCN), a small group of cells in the hypothalamus that generates a circadian rhythm reset by light entering the eyes. A rhythm is produced from the interaction of several proteins producing a biological clock.

Research

- Ralph et al. (1990) took the SCN out of genetically abnormal hamsters with a circadian cycle of only 20 hours, transplanting them into hamsters with the usual 24-hour cycle and their cycle shortened to 20 hours, suggesting that the SCN is the main endogenous pacemaker.
- Morgan (1995) found that removing the SCN from hamsters caused their circadian rhythm to disappear, but when SCN cells were transplanted in, the rhythm returned, again showing the central role of the SCN as an endogenous pacemaker.

- Hawkins and Armstrong-Esther (1978) found that shift-work altered nurses' sleep–wake cycles, but not their temperature cycles, suggesting that different body clocks regulate different circadian rhythms.

Role of exogenous zeitgebers

Light is seen as the most important zeitgeber, with the moon, the seasons, weather patterns and food availability among other ones. Zeitgebers play an important role in regulating biological rhythms, helping to reset them, and endogenous pacemakers need to respond to zeitgebers, coordinating the behaviours they regulate with the external environment.

Research

- Klein et al. (1993) studied a blind man with a circadian rhythm of 24.5 hours, which eventually got out of sync with the 24-hour day. Time cues such as clocks did not help and he had to take stimulants and sedatives to regulate his sleep–wake cycle, suggesting that light acts as an exogenous zeitgeber in the form of a time cue.
- Luce and Segal (1966) found contradictory results, as people in the Arctic Circle sleep for 7 hours even though it is light all summer long, implying that social cues act as zeitgebers here.

  Evaluation

- There is an adaptive advantage in animals having endogenous pacemakers reset by exogenous zeitgebers, keeping them in tune with seasonal changes, day/night changes, etc.
- To rely solely upon exogenous zeitgebers could threaten survival, so internal cues are also important.

Consequences of disrupting biological rhythms

Usually exogenous zeitgebers change gradually, giving time to adjust. However, rapid change disrupts coordination between internally regulated rhythms and external exogenous zeitgebers, creating consequences for our ability to function properly.

Jet lag

Jet lag is caused by travelling across time zones so quickly that biological rhythms do not match external cues, causing sleepiness during the day and restlessness at night. This lasts until resynchronisation has occurred; best achieved by being allowed to follow exogenous zeitgebers (e.g. staying awake until night-time). Jet lag is worse travelling west to east, as it is easier to adjust biological clocks if they are ahead of local time (**phase delay**) rather than behind (**phase advance**). Another reason for jet lag is that the biological clock regulating temperature needs time to reset, causing desynchronised rhythms in the meantime.

Research

- Klein et al. (1972) tested eight participants flying between the USA and Germany. They found that adjustment to jet lag was easier for people on westbound flights than eastbound, regardless of whether on an outbound or homebound flight, supporting the notion that phase advance has more severe consequences.

- Schwartz et al. (1995) found that baseball sides from the east of the USA played better against teams in the west than western sides did playing against teams in the east, suggesting that phase advance has severer consequences. However, it may be that eastern teams are superior.
- Webb and Agnew (1971) found that successful strategies for coping with jet lag include: outdoor pursuits, exposure to light and regular meal times. This suggests that following exogenous zeitgebers is the best way to address the consequences of jet lag.

Shift-work

Shift-work can involve working at times when normally asleep and therefore being asleep at times normally awake, causing breakdowns in the usual coordination between internal biological clocks and external cues. Many shift-workers change their working hours every week, causing severe disruption to normal routines of eating, resting, etc. Workers can be in an almost permanent state of desynchronisation, impairing concentration and physical performance and increasing stress levels that incur long-term health risks. Research into shift-work patterns shows that changing shifts forward in time causes less disruption, as does an adjustment time before changing shifts.

Research

- Czeisler et al. (1982) studied shift-workers at a factory in Utah, finding that they had high illness rates, sleep disorders and elevated levels of stress, suggesting that their internal body clocks were out of synchronisation with exogenous zeitgebers. He persuaded management to move to a phase delay system of rotating shifts forward in time, to reduce negative effects. Shift rotations were adjusted to every 21, instead of 7 days, giving time for adjustment. Nine months later, workers appeared healthier and more content, and output was up, showing how psychological research can lead to practical applications incurring positive outcomes.
- Sharkey (2001) reported on the beneficial effects of melatonin in reducing the time required to adjust to shift-work patterns and rotations, again demonstrating the positive effects of practical applications based on psychological knowledge.
- Hawkins and Armstrong-Esther (1978) studied nurses working night shifts. They found that their performance improved over a week, showing that the circadian rhythm adjusts gradually. However, their body temperature regulation was still desynchronised, suggesting that the temperature body clock takes longer to adjust.
- Monk and Falkard (1983) studied two types of shift: **rapidly rotating**, where shifts rotate quickly, and **slowly rotating**, where shifts rotate over a longer time period. Negative consequences were significantly more noticeable with rapidly rotating shifts, suggesting that they are more disruptive as they do not allow time for biological adjustments

- Colligan et al. (1978) found that workers with shift rotations had more accidents than workers on set shifts. They also drank more alcohol, took more sleeping tablets, and had digestive disorders, colds, anxiety, tiredness and less successful social relationships, demonstrating the destructive consequences of disrupting biological rhythms.

Evaluation

- As our endogenous cycle is roughly 25 hours, it is easier to deal with phase delay than phase advance.
- Research helping to understand the consequences of disrupting biological rhythms has practical applications, such as melatonin supplements to address jet lag.
- Research suggests that disrupting biological rhythms affects cognitive and emotional functioning as well as physical functioning, demonstrating the severity of consequences.
- There are large individual differences in the ways in which people are affected by shift-work and jet lag.
- A lot of research utilises naturalistic field studies. These are high in ecological validity, but incur many confounding variables, making the establishment of causality problematic.
- Serious incidents, like the Three Mile Nuclear Plant accident of 1979 and the Chernobyl reactor meltdown of 1986, occurred due to concentration and decision failures in the early hours of the morning. This suggests that the desynchronisation effects of working at irregular hours impair performance with disastrous consequences.
- The fact that concentration levels are affected by jet lag and shift-work suggests that they also create disruption to cognitive processes.
- The introduction of modern travel systems and electrical lighting has created a world that our biology, determined by evolution, cannot cope with, leading to disruption and negative consequences.

Sleep states

Specification

- *Nature of sleep*
- *Functions of sleep, including evolutionary explanations and restoration theory*
- *Lifespan changes in sleep*

To understand the nature of sleep, you need knowledge of the stages of sleep and the physiology of sleep.

Two specific theories are named as explanations of sleep and must be studied, as they could form specific examination questions. You also need knowledge of lifespan changes in sleep, from early infancy, through to old age. You are expected to be able to describe and evaluate all these features.

Nature of sleep

Sleep is a different state of consciousness where responsiveness to the external environment is diminished. It occurs daily as a circadian rhythm and is composed of an ultradian cycle of separate stages. With the invention of the electroencephalograph, psychologists were able to investigate brain activity occurring during sleep and concluded that it was composed of identifiably different sequential stages.

- Stage 1: alpha waves disappear and are replaced by low-voltage slow waves. Heart rate declines and muscles relax. This is a light sleep and the sleeper can easily be woken.
- Stage 2: a deeper state, in which the sleeper is still easily woken. Short bursts of sleep spindles are noticeable, together with sharp rises and falls in amplitude, known as K-complexes.
- Stage 3: sleep becomes increasingly deep, and the sleeper difficult to waken. Sleep spindles decline, being replaced by long, slow delta waves. Heart rate, blood pressure and temperature decline.
- Stage 4: deep sleep, where delta waves increase and metabolic rate is low. The sleeper is difficult to awaken. Growth hormones are released and incidences of sleepwalking and night terrors may occur.

The sleeper spends about 30 minutes in stage 4 sleep, with about an hour passing in total from stage 1 to stage 4. Stage 3 is re-entered, then stage 2 and then the sleeper enters an active stage of sleep called rapid eye movement (REM) about 90 minutes after falling asleep.

- REM: eye movements are noticeable, heart rate, respiration, etc. increase, and dreaming occurs.

After 15 minutes of REM sleep, the sleeper re-enters stages 2, 3 and 4 in that order, then another cycle begins. It is common to go through about five ultradian cycles in one night. As the night progresses, the sleeper spends more time in REM sleep and less time in other stages. This pattern is fairly universal, although there are developmental differences (see lifespan changes, p. 21).

Physiology of sleep

The brain stem has a role in key functions, such as alertness and arousal, but also controls sleep behaviour, with several hormones also being involved.

The SCN reacts to different levels of light received by the eyes, stimulating the production of **melatonin** from the pineal gland, which then stimulates the release of **serotonin** in the reticular activating system (RAS). The increase in serotonin levels causes RAS activity to lessen, bringing on the onset of sleep.

The release of **noradrenaline** causes the onset of REM sleep.

Another hormone, **acetylcholine**, is involved with brain activation during wakefulness and REM sleep, which is sometimes referred to as wakeful sleep.

Research
- Aserinsky and Kleitman (1955) found a relationship between dreaming and REM sleep, by waking people during periods of rapid eye movements and finding they were dreaming. However, this is reliant on subjective reports.
- Dement and Kleitman (1957) used EEG readings to monitor brain wave activity during uninterrupted sleep. They found that sleep consisted of a sequential series of five stages, each with common characteristics, occurring in a set pattern. Participants were woken at various times and reported dream activity mainly during REM sleep.
- Moor-Ede and Czeisler (1984) found that sleep occurs during the low point of our temperature cycle, showing the circadian nature of sleep as part of the daily sleep–wake cycle.

Evaluation
- The development of EEG readings gave psychologists an objective means of studying sleep behaviour.
- The majority of sleep studies occur in laboratories, with participants wearing electrodes, etc. They may therefore not reflect normal sleep patterns.
- Dement and Kleitman's study is not representative, involving few participants. However, similar studies show their findings to be reliable and valid.
- As sleep has five stages, it is likely that each stage has a different function. Because REM sleep is identifiable in warm-blooded creatures, but not cold-blooded ones, it might be that REM sleep serves the function, by increasing brain metabolism, of keeping brain temperature at a safe level.

Functions of sleep

Theories of sleep incorporate biological and psychological factors. Humans spend about a third of their time asleep, suggesting a crucial biological function. There is no single explanation for sleep, and all explanations have their weaknesses and strengths. Good theories should explain the universal nature of sleep, it being found throughout the animal world. Although there are wide variations between humans in sleep duration, sleep is generally seen as necessary, with the average being between 6 and 8 hours a night.

Evolutionary explanations

Evolutionary explanations see sleep as serving some adaptive advantage and occurring through natural selection. Different species evolved different types of sleep pattern, dealing with different environmental needs, such as predator avoidance, conservation of energy and dietary requirements. Sleep keeps animals dormant when activities vital for survival are not required.
- **Predator–prey sleep.** Meddis (1979) believes that sleep evolved to keep animals safely hidden from predators when usual activities, such as foraging, are not required. Therefore prey animals should sleep less, being more at risk and vigilant.
- **Hibernation theory.** Webb (1982) believes that active animals need larger amounts of food, threatening survival during times of food scarcity. Hibernation

conserves energy and increases survival. Grizzly bears hibernate through the winter, living off body fat accumulated during times of food availability.

- **Aquatic mammals.** The precise environmental demands of species affect sleep patterns and behaviours. Aquatic mammals need to breathe, so sleep incurs a risk of drowning. Animals have evolved strategies to cope with this problem.
- **Foraging needs.** Evolutionary explanations see sleep duration as affected by the amount of time needed to eat. Grazing animals spend a long time feeding, while predators can sleep a lot, only needing to eat periodically.
- **Body size.** Smaller animals evolved a greater need to sleep, their metabolic rates being high and energy consumption rapid. Long periods of sleep help to conserve energy stores.

Research

- Stear (2005) reported that sleep saves energy, keeps individuals from being lively at unnecessary times and is an adaptation to ecological factors differing across species, supporting the evolutionary basis for sleep.
- Requadt (2006) found that animals find warm, safe places to sleep as it minimises energy requirements to maintain body temperature. This supports the evolutionary point of view.
- Siegel (2008) reported that there is less risk of injury when asleep than awake, sleep being a safety device when essential activities are not necessary.
- Pilleri (1979) found that Indus dolphins sleep for a few seconds repeatedly, supporting evolutionary predictions for sleep patterns in aquatic mammals.
- Mukhametov (1984) found that bottlenose dolphins have one cerebral hemisphere asleep at a time, allowing animals to be asleep, alert and breathing simultaneously. This supports evolutionary predictions for the sleep patterns of aquatic mammals.

Evaluation

- Predators, such as lions, sleep longer than prey animals, such as zebras, seeming to support the evolutionary prediction. However, prey animals are usually herbivores needing time to graze, and therefore have less time to sleep.
- Giant sloths sleep for 20 hours a day, going against the evolution and body size argument, being large, inactive creatures with relatively low metabolic rates.
- Sleep may have evolved to suit human ecological needs in the Environment of Evolutionary Adaptiveness (EEA) but has little purpose in the modern world. However, it is still apparent.
- The fact that sleep is universal to species suggests some adaptive function.
- Sleep is so maladaptive in survival terms that it is difficult to see why it evolved. It prevents eating and reproduction, and incurs vulnerability to attack. However, the consequences of sleep deprivation can be severe.
- Evolutionary explanations are accused of reductionism, reducing complex behaviours down to adaptiveness. They are also deterministic, seeing behaviour as caused by past environments with no role for free will.
- A lot of research involving evolutionary explanations depends on animal studies, incurring a problem with generalising findings to humans.

Restoration theory

People generally sleep because they are tired, suggesting that sleep is fundamentally a period for rejuvenation and repair. Growth hormone is released during sleep, stimulating tissue growth and aiding protein synthesis, used to repair damaged tissues. Waste products are also removed.

Core sleep model

In the core sleep model (Horne 1988), stage 4 and REM sleep are seen as necessary for the healthy brain functioning required in cognitive processing. During these stages, the brain refreshes and restores itself ready for the challenges of the new day.

Brain and body restoration

According to restoration theory (Oswald 1980), high levels of brain activity during REM sleep indicate brain restoration, while growth hormone production (and other hormonal activity) during the four stages of slow-wave sleep (SWS) indicate bodily restoration and repair.

Research

- Stern and Morgane (1974) believe that during REM sleep, neurotransmitter levels are replenished, supporting the idea of restorative sleep. This is backed up by the fact that anti-depressants increase neurotransmitter levels, reducing REM activity.
- Cirelli et al. (2004) found that during SWS, genes associated with the protein production that regulates synaptic connections are activated, supporting the restoration theories, especially Oswald's.
- Horne (1988) reports that amino acids, built into proteins, are only available for 5 hours after eating. This suggests that protein synthesis cannot occur during sleep, casting doubts on to the theory.
- Shapiro et al. (1985) found that ultra-distance athletes running a double marathon slept for longer, suggesting that sleep does aid restoration.
- Everson et al. (1989) found that depriving rats of sleep causes increased metabolic rate, loss of weight and death in about 19 days, possibly due to immune system damage. This suggests that sleep is necessary for restoration. However, the rats are stressed to keep them awake and this may also contribute.
- Horne (1988) performed a meta-analysis of sleep deprivation studies. He found little evidence of reduced physical functioning or stress responses, suggesting that sleep is not primarily for restoration.

- Endurance-based athletes use short sleep sessions, after intensive training, to promote protein synthesis that repairs tissues, lending support to the idea of sleep being for restoration.
- Young infants have lots of sleep, possibly because of rapid brain and body tissue growth, supporting the idea of sleep for restoration.
- Fatal familial insomnia is a rare human condition, usually starting in middle age, where sufferers cannot sleep and die usually within 2 years, suggesting support for the

restoration theory. But cases are few and difficult to generalise from, sufferers also having brain damage that may be responsible.
- Several studies into physical exhaustion and sleep, such as Horne and Minard (1985), have found that participants fall asleep quicker, but not for longer, contradicting the findings of Shapiro et al.

Lifespan changes in sleep

There are wide differences in how much sleep individuals need, but there are important developmental changes, which within broad age groups are remarkably similar.
- **Neonates** sleep for about 16 hours a day over several sleep periods. After birth, infants display active sleep, an immature form of REM; this decreases and the amount of quiet sleep, an immature form of SWS, increases.
- **1-year-olds** develop brain activity sleep patterns increasingly like those of adults. The proportion of REM sleep declines to about 50% of sleep duration and sleep periods become longer and fewer. Total sleep is about 11 hours a day.
- In **5-year-olds**, brain activity during sleep resembles that of adults, although the frequency is different. REM sleep accounts for about a third of the total, which is about 10 hours daily, with boys sleeping slightly longer. Sleep disorders may be apparent, such as sleepwalking.
- **Adolescents** have less REM sleep than children and total sleep duration is now about 8–9 hours a night. Nocturnal sexual orgasms can occur.
- In **middle age**, normal adult sleep patterns are noticeable, but with increased levels of sleep disorders such as sleeplessness (insomnia) and snoring.
- In **senescence**, total sleep duration is unchanged, but REM sleep decreases to around 20% of the total, with stage 2 sleep increasing to about 60%. SWS decreases to around 5% and is non-existent in some cases. Sleep disturbance is common.

Research
- Van Cauter et al. (2000) examined several sleep studies involving male participants. Sleep was found to decrease during two life periods: between 16 and 25, and between 35 and 50.
- Floyd et al. (2007) reviewed nearly 400 sleep studies and found REM sleep decreasing by about 0.6% per decade. Its proportion increases from about age 70, though this may be due to overall sleep duration declining.
- Eaton-Evans and Dugdale (1988) found that the number of sleep periods for a baby decreases until about 6 months of age, then increases until 9 months of age, then slowly decreases again. This may be due to teething problems.

Evaluation
- Neonates sleeping for long periods may be an adaptive response, freeing up essential time for their parents.
- The fact that males over 45 years old can have little SWS, which affects hormone production, may explain why their physical injuries take longer to heal.

- During senescence the decline of SWS incurs lower levels of growth hormone production, suggesting that SWS is associated with its production.

Disorders of sleep

Specification

- *Explanations for insomnia, including primary and secondary insomnia and factors influencing insomnia, for example apnoea, personality*
- *Explanations for other sleep disorders, including sleepwalking and narcolepsy*

There is an explicit requirement to have knowledge of primary and secondary insomnias. Both are specifically named and could therefore form the focus of examination questions. Other factors need to be included, as the ones mentioned are examples and therefore not explicit requirements.

There is also a requirement to have knowledge of explanations for other sleep disorders, including sleepwalking and narcolepsy, which are named and could feature in the wording of examination questions. You should be able to describe and evaluate all of these features.

Explanations for insomnia

Insomnia is a sleep disorder where sufferers have long-term problems initiating or maintaining sleep. It can take the form of an inadequate quantity or quality of sleep. It is estimated that half of adults have problems with insomnia, with women suffering more than men, possibly because of hormonal fluctuations associated with onset of menstruation and the menopause. Tiredness is experienced during the day, affecting physical and cognitive functioning.

Primary insomnia

One in three insomniacs are primary insomniacs, where there are clearly apparent underlying causes. This could be a brain abnormality affecting the neural circuits involving sleeping; or environmental stress, which can be self-perpetuating, as not sleeping causes further stress, leading to a continuation of the insomnia and so on. Other common causes can be behaviour before sleep, sleep patterns and the sleeping environment: for instance, being too hot, cold or noisy.

Secondary insomnia

Secondary insomnia is where the disorder is secondary to some other medical condition. Psychological problems, such as depression, grief and dementia, account for around 50% of cases, with physical disorders, such as arthritis, diabetes and pain, accounting for another 10%. Medicines can also cause secondary insomnia, as can alcohol, caffeine and recreational drugs. Other sleep disorders can also incur insomnia.

Duration

Insomnia can be **transient**, lasting a few nights; **short-term**, lasting more than a few nights, but less than 3 weeks; or **long-term** (chronic), occurring most nights and for longer than 3 weeks.

Patterns

Onset insomnia involves difficulty in getting to sleep, often associated with anxiety. **Middle-of-the-night insomnia** is characterised by problems in getting back to sleep after waking, or waking too early. **Middle insomnia** involves waking in the middle of the night and/or difficulty in staying asleep, often associated with medical illnesses or physical pain. **Late insomnia** involves waking early in the morning, often associated with clinical depression.

Factors influencing insomnia

Apnoea

Apnoea is a medical condition where sufferers can have persistent pauses in their breathing lasting for minutes, as well as occasional loud snorts as breathing recommences. This can occur up to 200 times a night. **Obstructive sleep apnoea** is generally caused by blockage of the airways, and is often found in overweight middle-aged males. **Central sleep apnoea** occurs due to impaired brain signals to areas associated with breathing, and it happens more infrequently.

Research

- Chest (2001) found a significant positive correlation between insomnia and obstructive sleep apnoea, suggesting a relationship between the two conditions.
- Smith et al. (2004) found a relationship between insomnia and obstructive sleep apnoea, with sufferers being more prone to depression, anxiety and stress than people suffering from obstructive sleep apnoea, but not insomnia. Although this suggests a link between the two conditions, it seems that insomnia has more serious side-effects.
- Morrell et al. (2000) found that sleep apnoea is more common in older adults, with up to one in five sufferers — ten times the number of younger people — though the disorder tends to be more severe in the young. It was concluded that the difference in prevalence rates is due to changes in the structure and function of the cardiovascular system in older adults.
- Stickgold (2009) believes that a range of mental disorders, including depression and attention deficit disorder, may be caused by sleep apnoea and insomnia. It was found that apnoeic insomniacs had twice the incidence of depression than the normal population, suggesting that the best way to treat such mental disorders is to alleviate apnoea and insomnia.

Evaluation

- Sleep apnoea can lead to insomnia, which is more prevalent in older adults. As the population ages, there is an assumption that the disorder will grow, increasing the need for successful treatments.

- The higher incidence of apnoea in older adults is associated with changes to the cardiovascular system as we age, suggesting that sufferers of different ages require different treatments to address the problem.
- Doctors report increasing numbers of younger people with insomnia and sleep apnoea. This may be related to growing obesity among the young, suggesting that addressing obesity may be the best way to combat these sleep disorders.
- Horne (2009) points out that the claim of apnoea and insomnia causing mental disorders is not proven and that it is more probable that certain mental disorders lead to insomnia. He also believes obesity is more responsible for rises in depression levels.
- Research into apnoea and insomnia has led to the development of successful treatments, such as losing weight, stopping smoking, reducing alcohol and sleeping tablets intake, sleeping on your side and taking up exercise.

Circadian rhythm disruption

Disruption to the circadian rhythm through, for example, jet lag and shift-work can lead to insomnia (see 'Consequences of disrupting biological rhythms', p. 14).

Personality

Personality is implicated in research findings as being associated with the onset and continuation of insomnia. Psychasthenia, a personality disorder similar to obsessive–compulsive disorder, where a sufferer is plagued with unreasonable fears and doubts, excessive anxiety and obsessive compulsions, is especially implicated. Other commonly identified factors include over-sensitivity, low self-esteem, lack of autonomy and heightened emotional arousal.

Research

- Lundh et al. (1995) personality tested 233 persistent insomnia patients, finding that the predominant factor is psychasthenia, with high scores also on anxiety and monotony avoidance. Sufferers tend to score poorly on self-esteem and are overdependent on others. The effects of these factors are difficulty in regaining lost sleep, daytime fatigue and lack of concentration, suggesting that personality factors play a role in causing and maintaining insomnia.
- Kales et al. (1976) personality tested 124 insomniacs, finding that 85% had abnormal personalities characterised by psychasthenia, elevated levels of depression and conversion hysteria. Sufferers tended to internalise psychological disturbances, producing constant emotional arousal. This suggests that a psychophysiological mechanism underpins insomnia.
- Lahmeyer et al. (1989) found that insomniacs tend to have borderline abnormal personalities, characterised by traits such as manic episodes. This supports the idea that abnormal traits producing emotional arousal are associated with insomnia.
- De Carvalho et al. (2003) studied 32 Brazilian insomnia patients, finding heightened levels of anxiety and insecurity, especially among female patients.
- Grano et al. (2006) found that male insomniacs tend to be impulsive characters, implying that male and female sufferers are affected by different personality traits.

Evaluation

- Rather than personality traits leading to insomnia, there is a possibility that being insomniac creates changes in personality. Longitudinal studies following people at risk from their personality profile would be a good means of settling the debate.
- If it is proven that certain personalities can lead to insomnia, this suggests a practical application. At-risk people could be identified by personality testing and targeted for help and advice as a preventative means of treatment.
- Research suggests that treating abnormal personality traits and disorders has more success in reducing insomnia than treating insomnia to try and address personality defects. This implies that abnormal personality traits are the causal factor.
- Furukawa (2009) reports success in using behavioural treatments to address personality-linked insomnia, suggesting that maladaptive learning experiences may be important factors.

Explanations for other sleep disorders

Sleepwalking

Sleepwalking — known as **somnambulism** — refers to activities occurring unconsciously when asleep that normally occur when awake: for example, eating, cleaning and walking, and even sending nonsensical e-mails, having sex and committing murder. It is more prevalent in childhood, declining sharply in adulthood. Somnambulism tends to be associated with personality disorders, especially ones relating to anxiety.

Incidents of somnambulism occur during non-rapid eye movement (NREM) with the sufferer's eyes open, seemingly awake. However, speech tends to be gibberish and the episode is not recalled. The disorder places the sufferer at risk of injury or abuse.

Research

- Hublin et al. (1997) found that the disorder is more common among children, with up to 20% being affected. In adults the prevalence rate is about 2%. This indicates that the condition is linked to development and maturation.
- Broughton (1968) found the disorder to be heritable, with sufferers ten times more likely than the general population to have a close relative with the disorder, suggesting a genetic factor.
- Lecendreux et al. (2003) found a higher incidence among MZ (identical) twins than DZ (non-identical) twins: 50% compared to 12%, suggesting a genetic link, although identical twins also tend to have identical environments, which could be a confounding variable.
- Lecendreux et al. (2003) identified the DQB1*05 gene as being linked to somnambulism. However, it has also been linked to another sleep disorder, **sleep terrors**, occurring mainly in children
- Kales and Kales (1975) found that somnambulistic episodes generally occur in stage 3 and 4 sleep.
- Shapiro et al. (1996) reported cases of *sexomnia* where sufferers have somnambulistic sex. As with other forms of somnambulism, it occurs during

NREM sleep. Confirmation was made by monitoring behaviour with cameras and recording brain wave patterns, which showed unusual activity levels.

- Nowak (2004) reported on the case of an Australian woman who regularly left her house to have somnambulistic sex with strangers. She had a history of sleepwalking and sleeptalking since she was a child.

Evaluation

- It has been suggested, and is the current focus of research, that a group of sleep disorders known as parasomnias, such as somnambulism, sleep terrors, etc., share a common genetic cause.
- Children may have higher incidences of somnambulism because they spend long amounts of time in slow wave sleep, where the condition tends to occur. However, children tend to be observed more while asleep, so their episodes are more likely to be detected.
- Somnambulism has been used as a legal defence against charges of murder. Kenneth Parks was acquitted in 1987 of killing his in-laws in Canada due to non-insane automatism.
- Somnambulism has been alleviated by avoiding risk factors, such as excitatory activities, using techniques such as meditation before sleeping and sleeping in a safe environment.

Narcolepsy

Narcolepsy means 'seized by sleepiness' and is a sleep disorder characterised by disruption to the sleep–wake cycle, whereby sufferers suddenly fall asleep at unexpected times, often in the middle of activities. Sufferers feel sleepy during the day, appearing drunk. Sporadic microsleeps are common, from which the sufferer awakes without realising they have been asleep.

Another common symptom is **cataplexy**, where muscular control is lost, usually as a result of being aroused: for example, while excited. Sufferers can also suffer from **sleep paralysis**, occurring when the brain awakes from an REM state, leaving the body paralysed, with the sufferer fully conscious. Terrifying hallucinations, **hypnopompic** on awakening, or **hypnagogic** on falling asleep, and a general sense of danger, can accompany sleep paralysis.

The experience of narcolepsy is often perceived as a dream, with dream-like objects appearing alongside objects in vision. Sleep paralysis can last from a few seconds to several minutes, and is then followed by sensations of panic and a gradual realisation that the hallucinations were not real. Sufferers often wake frequently during proper night-time sleep too.

The condition usually appears in adolescence and is believed to be the result of a genetic abnormality. Other possible causes may be a shortage of the neurotransmitter **hypocretin**, or an autoimmune disease.

Not all cases are reported, as some are mild afflictions. It is estimated that 0.05% of people in Europe and the USA are sufferers, rising to 0.16% in Japan.

Research

- Hufford (1982) reports on an ancient sleep paralysis myth, hag riding. This offers an explanation for people's belief in witches, but also suggests that the condition is long lasting, implying that it may be hereditary. This is backed up by such legends being cross-cultural: Kanashibari (1993) reports similar myths in Japan.
- Montplaisir (2007) tested 16 patients with narcolepsy and cataplexy, finding that they had a higher percentage of REM sleep, although this may be a cause or an effect of the condition. Evidence was found of decreased hypocretinergic and/or dopaminergic abnormalities in input to brainstem structures, suggesting that abnormal levels of neurotransmitters are associated with the condition.
- Broughton (1999) reported that symptoms cannot be cured at present, but can be controlled with behavioural treatments and medication. Stimulants such as Modafinil combat daytime drowsiness, while cataplexy is addressed with anti-depressants, but drug treatments only lessen symptoms at best. Lifestyle adjustments, such as regulated sleep schedules and non-excitatory activities before bedtime, are more successful.
- Daniels et al. (2001) gave questionnaires to 500 patients of the Narcolepsy Association, finding that they had lower energy/vitality levels, reduced social functioning and lessened physical activity, with 57% of them depressed. The condition put limitations on their educational, work, home and social life, matching results from other countries. This demonstrated the extensive impact of the disorder on health-related quality of life.
- Mignot et al. (1999) used positional cloning to pinpoint a defective gene, hypocretin receptor 2, in dogs, one of the few species suffering from narcolepsy. The defective form of the gene encodes proteins that cannot recognise important signals, cutting off cells from receiving essential directives, including messages that promote wakefulness. This suggests a genetic basis to the disorder, although there is a problem in generalising to humans. However, the gene does occur in humans.

Evaluation

- The frightening and bizarre hallucinations that are part of the condition have been suggested by Blackmore and Cox (2008) as explaining the experience of alien abduction that people sometimes report.
- Using samples based on patient associations can cause problems with bias, as only certain types of sufferer may join the association.
- The identification of genes associated with the disorder does not mean that there is a definite genetic cause and researchers stress the need to identify environmental triggers. Genes need specific environments in which to express themselves.
- Genetic research may lead to the production of drugs compensating for the failure of the hypocretin system. Such drugs may work better than the current regime of stimulants and anti-depressants.
- It is hoped that research into narcolepsy may help create sleeping drugs more closely mimicking natural brain chemistry.

Relationships

Formation, maintenance and breakdown of romantic relationships

Specification

- *Theories of the formation, maintenance and breakdown of romantic relationships: for example, reward/need satisfaction, social exchange theory*

The focus is on romantic relationships, with the subject matter breaking down into three separate, but interrelated areas: formation, maintenance and breakdown. No specific theories are named; the two given are examples and any other relevant ones would suffice, as examination questions will not be directed at explicitly named theories. You should be able to describe and evaluate theories.

Theories could relate to more than one feature of relationships (i.e. to formation, maintenance or breakdown), so it is important to shape the use of material to fit the demands of the question being answered.

An acceptable strategy is to put specific theories together to form general theories: for example, combining social exchange theory and equity theory, or the reinforcement-affect model with need satisfaction.

Formation of romantic relationships

People are attracted to and form relationships with others for many reasons. These can be highly individual, but attempts have been made to formulate general theories explaining the various phases of romantic relationships and to take into account cultural and gender influences.

Sociobiological explanation

The sociobiological explanation is an evolutionary theory perceiving relationship formation as a form of 'survival efficiency', with a different focus between genders. Males are never certain of paternity and produce vast amounts of sperm — their best strategy to further their genes being to have multiple partners. Males look for signs of fertility, such as smooth skin, and sexual faithfulness, as they do not want to waste resources bringing up another male's child.

Females produce a small number of eggs, but are certain of the maternity of children. Females seek to ensure that their children are genetically strong and healthy by being selective in choosing partners and getting them to invest resources. Females look for kindness, which indicates a willingness to share resources. The more a male invests, the more likely he will not desert and will offer further resources.

Males compete to be selected and females choose based upon characteristics reflecting genetic fitness. Courtship serves as a period in which competition and selection occur and in which males are encouraged to invest resources, increasing the chances of them not deserting and investing more resources.

Research

- Dunbar (1995) analysed 900 personal advertisements and found that 42% of males sought youthfulness, while only 25% of females did; 44% of males sought attractiveness, while only 22% of females did. This supports the sociobiological idea that males and females have different reasons for forming relationships.
- Davis (1990) performed a content analysis of personal advertisements and found that men look for health and attractiveness, while offering wealth and resources. Females look for resources and status, while offering beauty and youth. These findings support the idea of gender differences in relationship formation based on evolution.
- Harris (2005) examined cultures dominated by different religious systems. It was found that relationship behavioural patterns either contradicted sociobiological strategies of relationship formation, or placed stress on cooperative restraint rather than survival through selfish propagation, which is predicted by sociobiological theory.
- Packer (1983) reported that male lions, upon defeating dominant males, kill the existing cubs, and the lionesses then become sexually receptive. This behaviour is adaptive, eliminating competition for the cubs and allowing faster genetic reproduction, supporting the sociobiological explanation.

Evaluation

- The explanation is reductionist, seeing relationships purely as a means of reproducing.
- It presumes heterosexuality, that children are wanted and that all relationships are sexual; it is therefore oversimplified.
- The explanation can be seen to support gender stereotypes of housebound women and sexually promiscuous males.
- The theory offers a plausible explanation for the evolution of mate preferences.
- Although relevant to the Environment of Evolutionary Adaptiveness, the explanation does not suit the modern environment. For example, many women now have resources.
- The explanation is deterministic, disregarding the role of free will in relationship formation.

Reinforcement and need satisfaction

This theory sees conditioning as an explanation for relationship formation. A person may reward us directly (operant conditioning) by meeting psychological needs for love and sex, or indirectly (classical conditioning) because they become associated with pleasant circumstances, therefore we are more likely to form a relationship. So if we associate a person with being in a good mood, or helping to remove a negative mood, we will find that person attractive, increasing the chances of relationship formation. Argyle (1994) outlined how forming a relationship can satisfy social needs on several levels: **biological needs** (such as collective eating), **dependency** (being comforted), **affiliation** (a sense of belonging), **dominance** (making decisions for others), **sex** (flirting), **aggression** (letting off steam) and **self-esteem** (being respected by others).

Research

- Cunningham (1988) studied males who watched a happy or sad film and then interacted with a female. More positive interactions came from those watching the happy film, supporting the explanation.
- May and Hamilton (1980) asked females to rate photos of males, while nice or unpleasant music was played. Those with the nice music rated the males as more attractive, supporting the theory.
- Griffit and Guay (1969) conducted a study in which participants were evaluated on a creative task by the experimenter. If the evaluation was positive, participants expressed more liking for a non-involved bystander than if it was not positive, supporting the idea of people being liked who are associated with positive outcomes.
- Griffit and Veitch (1971) found that evaluations of a stranger were positive when the evaluation was made in comfortable surroundings, supporting the theory.
- Hays (1985) investigated student friendships and found that rather than just being focused on rewards received, individuals favoured equity, giving priority to rewarding the other person too, weakening the explanation.

Evaluation

- A lot of research is laboratory based and lacks ecological validity, as tasks carried out are not usual ones undertaken and do not reflect real relationships. Therefore they do not add to understanding of relationship formation.
- The explanation is reductionist, seeing only rewards and need satisfaction as being involved in relationship formation, and therefore neglecting other factors.
- The explanation is deterministic and oversimplified, seeing relationship formation as an unconscious process based on learned associations, and therefore not allowing for free will and cognitive processing.
- Many non-western cultures feature relationships that do not have regard for receiving rewards or prioritising selfish needs. Therefore the theory cannot account for cultural differences.
- The explanation also cannot account for gender differences. Women often focus more on the needs of others, and males and females tend to find different things rewarding, suggesting that the explanation is oversimplified.

Maintenance of romantic relationships

Social exchange theory

Social exchange theory — also known as **economic theory** — explains relationships in terms of maximising benefits and minimising costs. The 'social exchange' is the mutual exchange of rewards between partners, such as friendship and sex, and the costs of being in the relationship, such as freedoms given up. A person assesses their rewards by making two comparisons:

- the **comparison level** (CL), where rewards are compared to costs to judge profits
- the **comparison level for alternative relationships** (Clalt), where rewards and costs are compared against perceived rewards and costs for possible alternative relationships

A relationship will be maintained if rewards exceed costs and the profit level is not exceeded by a possible alternative relationship.

Thibaut and Kelley et al. (1959) proposed a four-stage model, setting out how relationships could be maintained (see Table 1). It perceives that over time people develop a predictable and mutually beneficial pattern of exchanges, assisting the maintenance of relationships.

Table 1 Thibaut and Kelley's stages of relationship formation and maintenance

Stage	Description
Sampling	Rewards and costs are assessed in a number of relationships.
Bargaining	A relationship is 'costed out' and sources of profit and loss are identified.
Commitment	The relationship is established and maintained by a predictable exchange of rewards.
Institutionalisation	Interactions are established and the couple have 'settled down'.

Research

- Rusbult (1983) asked participants to complete questionnaires over a 7-month period concerning rewards and costs associated with relationships. It was found that social exchange theory did not explain the early 'honeymoon' phase of a relationship, where balance of exchanges was ignored. However, later on, relationship costs were compared to the degree of personal satisfaction, suggesting that the theory is best applied to the maintenance of relationships.
- Rusbult (1983) found that the costs and rewards of a relationship were compared to the costs and rewards of potential alternative relationships to decide if the relationship should be maintained, supporting the social exchange model.
- Hatfield (1979) looked at people who felt over- or under-benefited. The under-benefited felt angry and deprived, while the over-benefited felt guilty and uncomfortable, supporting the theory by suggesting that, regardless of whether individuals are benefited, they may not desire to maintain a relationship.

Evaluation

- The social exchange theory was modified into the equity theory, which concerns balance and stability in a relationship and can be regarded as a logical progression.
- Argyle (**1988**) criticised methodologies used to evaluate social exchange theory, declaring them contrived and artificial with little relevance to real life.

- Research has tended to concentrate on short-term consequences of relationships rather than more important long-term maintenance.
- The theory applies to people who 'keep score'. Murstein et al. (1977) devised the exchange orientation tool, identifying such scorekeepers. They were found to be suspicious and insecure, suggesting that the theory only suits relationships lacking confidence and mutual trust.

Equity theory

Equity theory perceives individuals as motivated to achieve fairness in their relationships and to feel dissatisfied with inequity (unfairness). Maintenance occurs through balance and stability. Relationships where individuals put in more than they receive, or receive more than they put in, are inequitable, leading to dissatisfaction and possible dissolution. Definitions of equity within a relationship can differ between individuals.

A relationship may alternate between periods of perceived balance and imbalance, with individuals being motivated to return to a state of equity. The greater the perceived imbalance, the greater the efforts to realign the relationship, so long as doing so is perceived to be viable.

Walster et al. (1978) saw equity as based on four principles, as shown in Table 2.

Table 2 Principles of equity (Walster et al. 1978)

Principle	Description
Profit	Rewards are maximised and costs minimised.
Distribution	Trade-offs and compensations are negotiated to achieve fairness in a relationship.
Dissatisfaction	The greater the degree of perceived unfairness, the greater the sense of dissatisfaction.
Realignment	If restoring equity is possible, maintenance will continue, with attempts made to realign equity.

Research

- Yum et al. (2009) looked at different types of heterosexual romantic relationship in six different cultures. As predicted by equity theory, maintenance strategies differed, with individuals in perceived equitable relationships engaging in most maintenance strategies, followed by those in perceived over-benefited and under-benefited relationships. Cultural factors had no significant effect on these findings, suggesting that equity theory can be applied to relationships across cultures.
- Canary and Stafford (1992) devised the 'Relationship Maintenance Strategies Measure' (RMSM), using it to assess the degree of equity in romantic relationships. A link was found between the degree of perceived equity and the prevalence of maintenance strategies, implying that equitable theories are maintained.
- Dainton (2003) studied 219 individuals in romantic relationships, finding that those in relationships of perceived inequity had low relationship satisfaction, but were

motivated to return to an equitable state in order to maintain the relationship. This suggests that equity is a main factor in relationship satisfaction and maintenance.

Evaluation

- Equity may be more important to females, suggesting that the theory is not applicable to both genders. Hoschchild and Machung (1989) found that women do most of the work in making relationships equitable.
- Sprecher (1986) believes that close relationships are too complex to allow precise assessment of various rewards and costs involved in establishing equity.
- Mills and Clark (1982) believe it is not possible to assess equity in loving relationships, as much input is emotional and therefore unquantifiable, and to do so would diminish the quality of love.
- Some research has suggested that equity theory does not apply to all cultures. Moghaddam et al. (1983) found that US students prefer equity, but European students prefer equality, suggesting that the theory is a reflection of the values of US society.

Breakdown of romantic relationships

Duck's four-stage theory

Duck (1984) proposed a four-stage theory of relationship dissolution, shown in Table 3.

Table 3 Four-stage theory of relationship dissolution (Duck 1981)

Phase of dissolution	Description
Intra-psychic phase	One partner privately perceives dissatisfaction with the relationship.
Dyadic phase	The dissatisfaction is discussed. If it is not resolved, there is a move to the next stage.
Social phase	The breakdown is made public. Negotiation about children, finances, etc. with wider families and friends becoming involved.
Grave-dressing phase	Establishing post-relationship view of the break-up, protecting self-esteem and rebuilding life towards new relationship.

Duck (1981) saw two types of cause for relationship dissolution:
- **predisposing personal factors**, such as an individual's bad habits or emotional stability
- **precipitating factors**, such as exterior influences (e.g. love rivals), process features (e.g. incompatible working hours), emergent properties (e.g. lack of relationship direction) and attributions of blame (e.g. perceiving someone else being to blame)

Duck also saw a **lack of skills** (e.g. being sexually inexperienced), a **lack of motivation** (e.g. perceiving inequity) and a **lack of maintenance** (e.g. spending a lot of time apart) as being important.

Research

- Kassin (1996) found that women are more likely to stress unhappiness and incompatibility as reasons for dissolution, while men blame lack of sex. Women wish to remain friends, while males want a clean break, suggesting gender differences that the model does not consider.
- Argyle (1988) found that women identified lack of emotional support as a reason for dissolution, while men cited absence of fun, again suggesting gender differences that the model does not explain.
- Akert (1992) found that the person who instigated the break-up suffers less negative consequences than the non-instigator, suggesting individual differences in the effects of dissolution that the model does not explain.

Evaluation

- There are cultural differences in relationship dissolution which the model does not explain. Many non-western cultures have arranged marriages, which can be more permanent and involve whole families in crises.
- The model does not apply to homosexual relationships, which do not involve decisions over children that heterosexuals have to consider.
- The model is simplistic as it does not account for a range of relationships, such as casual affairs and friendships.
- The model is plausible, relating to lots of people's experiences of relationship dissolution.
- The model has practical applications in counselling. Assessing which phase a couple are in can help form strategies to rescue relationships.
- The phases of breakdown are not necessarily universal — not all couples go through them or go through them in that particular order.

Lee's five-stage theory

Lee (1984) proposed a five-stage model of relationship dissolution, seeing dissolution as a process occurring over time, rather than just being a single event.

Stage of dissolution	Description
Dissatisfaction	An individual becomes dissatisfied with the relationship.
Exposure	Dissatisfaction is revealed to one's partner.
Negotiation	Discussion occurs over the nature of the dissatisfaction.
Resolution	Attempts are made to resolve the dissatisfaction.
Termination	If the dissatisfaction is not resolved, the relationship ends.

Research

- Lee created his theory after conducting a survey of 112 break-ups of non-marital romantic relationships, finding that the **negotiation** and **exposure** stages were most distressing and emotionally exhausting. Individuals who missed out stages, going straight to termination, tended to be those with less intimate relationships. Those going through the stages in lengthy and exhaustive fashion felt attracted to their former partner after termination and experienced greater feelings of loss and loneliness.

- Argyle and Henderson (1984) asked participants to consider whether rule violations were to blame for personal relationship breakdowns and, if so, which ones. Rule violations were found to be important factors, with jealousy, lack of tolerance for third-party relationships, disclosing confidences, not volunteering help and public criticism most critical, suggesting that Lee's explanation cannot be complete as it does not account for these factors.
- Research studies relating to Duck's model can also generally be applied to Lee's theory.

Evaluation

- Like Duck's, Lee's theory has practical applications in relationship counselling. For example, if a couple are in the exposure stage, then counsellors can concentrate on re-establishing affection in the relationship.
- Lee's theory is culturally specific, not relating to non-western cultures, such as collectivist cultures and those cultures with a tendency for arranged marriages.
- Lee's theory can be regarded as reductionist, focusing only on romantic, heterosexual relationships. This suggests that the theory is not applicable to friendships, homosexual relationships, etc.
- The theory is simplistic, as it cannot explain the whole range of relationships and reasons for dissolution.
- One strength of Lee's research is that a lot of information was gathered and the sample was large. However, it only contained students in pre-marital relationships and may not relate to other relationships, especially long-term relationships involving children and shared resources.
- Lee's theory is more positive than Duck's, seeing more opportunities for problematic relationships to be saved.
- The theory (like Duck's) cannot explain abusive relationships where the abused partner may not initiate the stages of dissolution, being reluctant to reveal their dissatisfaction. Instead the abused partner may simply walk away from the relationship.
- Stage theories describe the process of dissolution, but do not provide explanations of why the process occurs.

Human reproductive behaviour

Specification

- *Relationship between sexual selection and human reproductive behaviour*
- *Evolutionary explanations of parental investment: for example, sex differences, parent–offspring conflict*

The emphasis here is on human reproductive behaviour, so there is a need to explain sexual selection from the viewpoint of males and females. No definite theories or research content is specified, but there is a selection of appropriate material that can be utilised.

The specification then focuses upon evolutionary explanations of parental investment. The examples given would not feature directly in any examination question, so any relevant material would serve equally well.

Relationship between sexual selection and human reproductive behaviour

Sexual selection involves the selection of characteristics increasing reproductive success. For example, if a bird's plumage enhances his prospects of being chosen as a mate, the characteristic becomes enhanced as a sexually selected one.

Reproductive success involves the production of healthy children, who survive to sexual maturity and reproduce themselves.

There are differences between male and female sexual behaviour, as they are subject to different selective pressures. These differences occur due to **anisogamy** — differences between the nature and amount of gametes (sperm and eggs) produced.

Males produce a lot of small, highly mobile sperm and can fertilise many females at little cost to reproductive potential. They cannot be sure of paternity, so natural selection favours male behaviours maximising the number of potential pregnancies. This results in **intrasexual** competition between males and **polygamy**, a mating pattern in which one individual mates with more than one individual of the opposite sex. Males seek females displaying signs of fertility, such as health, youth and childbearing hips.

A female produces a few, relatively large eggs, each one representing a sizeable reproductive investment, although she is always sure of maternity. Natural selection favours female behaviours maximising chances of each potential reproduction being successful, such as careful mate selection, monogamy and high parental investment. Females seek males displaying genetic fitness, such as strength, status and resources. Females indulge in **intersexual** competition, where females choose males from those available.

Research
- Buss (1989) tested participants from 37 cultures, finding that males prefer young, physically attractive females, while females prefer resource-rich, ambitious, industrious males, supporting the gender-based predictions of sexual selection in humans.
- Dunbar and Waynforth (1995) looked at personal advertisements, finding that 42% of males, compared to 25% of females, sought youthful partners, while 44% of males and 22% of females sought physical attractiveness, supporting the prediction.
- Davis (1990) performed a content analysis of personal advertisements, finding that men look for health and attractiveness, while offering wealth and resources.

Females look for resources and status, while offering beauty and youth, supporting the idea of gender-based differences in sexual selection.

- Singh (1993) found that males prefer females with a waist:hip ratio of 0.7:1, suggesting choice on the basis of potential fertility.
- Clark and Hatfield (1989) found that males are more promiscuous, supporting the idea of gender-based differences in sexual selection.
- Boone (1986) found that females prefer older males with access to resources, while Kenrick and Keefe (1992) found that males prefer younger females, supporting the theory.
- Penton-Voak et al. (2001) found that females prefer males with greater facial symmetry, an indication of developmental stability that would be passed on to their sons, increasing reproductive potential.

Male strategies

It has been proposed that males adopt the following reproductive strategies:

- **Courtship rituals.** These allow for males to compete and display genetic potential, by their characteristics and resource abilities. Miller (1997) sees evolution as shaping human culture (i.e. language, art, humour and music) to act as courtship displays, attracting sexual partners.
- **Size.** Males have evolved to be bigger, demonstrating strength for success in competition against other males. Weaponry has evolved in some species for the same end (e.g. antlers).
- **Sperm competition.** Natural selection has acted upon males, making them more competitive by producing larger testicles, more copious ejaculations and faster-swimming sperm.
- **Jealousy.** Males fear being cuckolded and spending resources raising another male's child. Buss (1993) found that men were fearful of their partner being sexually unfaithful, while females were fearful of emotional unfaithfulness, illustrating the male fear of cuckoldry and the female fear of their partner spending resources on another female.
- **Sneak copulation.** Males will mate with females other than their partner if given the opportunity. Women gain from this, as having different fathers brings a wider genetic diversity to her children, increasing survival chances. Cerda-Flores (1999) found a non-paternity rate of 12% among children in Mexico.

Female strategies

The following hypotheses have been advanced for female reproductive strategies:

- **The sexy sons hypothesis.** Females select attractive males who will produce sons with the same attractive features, increasing reproductive fitness. Attractive characteristics have an initial adaptive advantage, but natural selection favours their enhancement, until they 'runaway', becoming bizarre, like the complex and highly decorated bowers that male bower birds construct to attract females.
- **The handicap hypothesis.** Zahavi (1975) believes that females select males with handicaps because it advertises their ability to thrive despite handicaps, demonstrating superior genetic quality. This may be why females find males who

can drink/take drugs in large amounts attractive as they are demonstrating their ability to handle toxins, a sign of genetic fitness.

Evaluation

- Males of polygamous species are under the greatest selective pressure to exhibit sexual display characteristics. Kirkpatrick (1987) found that this created a runaway process, often resulting in features maladaptive in circumstances other than courtship, supporting the idea of the handicap hypothesis.
- There is a difficulty in identifying and separating the effects of sexual selection from natural selection, making research into this area problematic.
- Partridge (1980) allowed some female fruit flies to mate freely, forcing others to mate with randomly chosen males. The offspring of these matings were then tested for competitive ability by being raised with a fixed number of standard competitors. The offspring of the free choice females did better, suggesting that females can improve the reproductive success of their children by selecting good genes in their partners.
- Care should be taken when generalising the findings of animal studies on to humans.
- Moller (1992) explained why females choose males with symmetrical features in terms of the handicap hypothesis, symmetry requiring genetic precision, which can only be produced by good genetic quality males.
- There are examples in the animal kingdom of sneaky copulations in apparently monogamous species. Birkhead (1990) found that 8% of Zebra finch offspring were the result of females' sneaky copulations with males other than their partner.
- There is evidence from the animal kingdom supporting the idea of sperm competition. Dewsbury (1984) reported that rats have a mating system where multiple males mate with a female, especially at high population densities. Rats have large testicles for their body size, allowing the production of copious sperm, increasing the chances of reproduction.
- Different studies into human sneak copulations found widely differing figures. Sasse et al. (1994) in a Swiss study found a low figure of 1.4%; other studies have suggested a rate as high as 20%. This may be due to cultural differences, or to types of sample used — for instance, using DNA data from examples where males had suspicions of non-paternity would be more likely to find such evidence.
- Evolution explains male and female behaviour in terms related to maximising reproductive potential. However, the same behaviours can be explained in other ways: for example, female choosiness and male promiscuity can be explained by gender role socialisation.

Evolutionary explanations of parental investment

Parental investment is investment increasing a child's survival chances at the expense of the parent's ability to invest in other children, either living or yet to be born.

A male's investment is smaller than a female's. The only restriction on his reproductive ability is how many females he can mate with. The male produces vast quantities of sperm over a long period of time, incurring few costs in the way of time or energy. He cannot be sure of paternity and his best strategy is to impregnate as many females as possible.

A female's investment is much larger. She produces relatively few eggs, is fertile for a far shorter period and bears the costs of pregnancy and those of childrearing. Her best strategy is to indulge in behaviours increasing the survival chances of her children.

Other explanations of parental investment are as follows:

- **Paternal certainty.** With internal fertilisation, males are more likely to desert than with external fertilisation, as they are unsure of the offspring's paternity.
- **Order of gamete release.** Internal fertilisation gives males the chance to desert and leave childcare duties to the female, while with external fertilisation the female has the opportunity to desert first.
- **Monogamy.** In species where offspring are born at an early stage of development or where childcare is intensive, pair bonds tend to be exclusive and long lasting, increasing the chances of the offspring's survival.
- **Parental certainty.** Maternal grandparents are certain a grandchild is genetically related to them, while paternal grandparents are not. Therefore more care and resource allocation will come from maternal grandparents than paternal ones.

Research

- Gross and Shine (1981) report that with internal fertilisation parental care is carried out by females in 86% of species, while with external fertilisation parental care is carried out by males in 70% of species, supporting the predictions based on paternal certainty.
- Krebs and Davies (1981) report that the males of some fish species release sperm first in a nest and then the female lays her eggs, meaning that the male has the first opportunity to desert, but the males carry out the childcare, going against the idea of the order of gamete release hypothesis.
- Daly (1979) reports that in some mammals and birds monogamy and bi-parental care is often apparent, because the males contribute to feeding and caring for the young, supporting the predictions for monogamy.
- Pollett et al. (2007) found that over 30% of maternal grandparents and 25% of maternal grandfathers had regular contact with their grandchildren, while only 15% of paternal grandparents did, fitting the predictions of parental certainty.
- Brase (2006) found that males who demonstrate cues of a positive disposition to parental investment were seen as more attractive by females, supporting the evolutionary explanation of parental investment.

Evaluation

- It is possible to test evolutionary theory by making predictions based on the theory and then seeing if real-life examples support the predictions.
- Krebs and Davies (1981) report that it is not always true that external fertilisation leads to increased paternal certainty. In sunfishes, cuckoldry occurs during the female's egg positioning.

- Dawkins and Carlisle (1976) found that in 36 out of 46 species where there is simultaneous gamete release and both sexes have equal chances of deserting, it is the males who provide monoparental care, refuting the prediction somewhat.
- Andersson et al. (1999) looked at investments by fathers in the college education of their biological and stepchildren, finding that they were highest when fathers were living with the biological mother of their children, but otherwise investments were equal, going against evolutionary theory. Maybe men invest in stepchildren to show their ability as a resource provider, increasing their attractiveness to females.

Parental–offspring conflict

The behaviour of offspring can influence parental investment. Parents have equal investments in all offspring, but the amount of resources allocated to each one decreases as more are born and as an individual child gets older and is able to fend for itself. However, individual offspring will try to get parents to invest more in them at the expense of any other offspring. This creates **sibling rivalry**, with offspring competing for attention and resources.

Parent–offspring rivalry can occur before birth, with a mother experiencing high blood pressure due to the foetus secreting hormones to gain more nutrition, the resulting high blood pressure bringing more nutrition to the foetus.

Children will use various strategies to try and manipulate parents into allocating them resources, such as crying, smiling and regression to an earlier state. These can occur as solitary acts, or ones performed at the expense of siblings' needs.

Older parents will often tolerate the demands of young infants more, as they are not compromising their future reproductive potential. If they are not having any more children, it makes sense to centre resources on the existing ones.

Research
- Rimm (2002) found sibling rivalry most intense when children are close in age and need resource investment more. Sibling rivalry was also intense when one sibling was gifted, presumably as an attempt to stop the talented sibling receiving advantageous proportions of parental resources.
- Sulloway (2001) found that, among spotted hyenas, sibling rivalry occurs as soon as a second cub is born, with 25% of cubs being killed by their siblings.
- Haig (1993) found that women experiencing high blood pressure have fewer spontaneous abortions, while Xiong et al. (2000) found they have larger babies, suggesting that high blood pressure has an adaptive advantage in producing healthier babies.
- Trivers (1985) found that herring gulls manipulate parents by appearing smaller than they are, to elicit feeding.
- Goodall (1990) reported that juvenile chimpanzees act in a babyish manner to get parental attention, suggesting that the juvenile is maximising the amount of resources it receives and maybe delaying the mother from breeding again, as this would be disadvantageous to the juvenile.

Evaluation

- Children have motivation to feel negatively towards siblings, as they are co-competitors for limited parental resources, but also are motivated to have positive regard for siblings as they share 50% genetic similarity, explaining the contradictory behaviour that siblings display to each other.
- Human parents often demonstrate a strategy to cope with sibling rivalry by taking them along different developmental paths, maximising each individual's strengths, reducing conflict and leading to very different individuals.
- The tendency for expectant mothers to feel nauseous is explained by Profet (1992) as being induced by the foetus to protect itself from poisonous substances in the mother's diet. Food cravings can be explained similarly as a foetus manipulating a mother's diet to its advantage.
- Children's temper tantrums can be seen in an evolutionary light as the child about to injure itself or attract the attention of predators. Thus parents attend to its demands to reduce the risk.
- Evolutionary theory is accused of being reductionist, as it reduces human parental behaviour down to the single explanation of adaptive fitness, thus ignoring other possible explanations. It is also accused of being deterministic in seeing parental behaviour as driven by biological factors with no input for free will.

Effects of early experience and culture on adult relationships

Specification

- *Influence of childhood and adolescent experiences on adult relationships, including parent–child relationships and interaction with peers*
- *Nature of relationships in different cultures*

The specification focuses here on childhood and adolescence experiences and their effects on adult relationships. Parent–child relationships and interaction with peers are explicitly named and could form the basis of an examination question, either singularly or in combination with each other.

There is also a requirement to have knowledge of the nature of relationships in different cultures, to such an extent that they can be both described and evaluated.

Influence of childhood experiences on adult relationships

Individuals differ in their relationships: some are content in long-term relationships, while others prefer more short-term, less intense associations. There are those who seem 'lucky in love', while others appear to lurch from one unsuitable relationship to the next. Psychologists have tried to see if the quality and pattern of relationships in adulthood is linked to earlier experiences.

Continuity hypothesis

Bowlby (1951) believed that the type and quality of relationship that an individual has with their primary caregiver provides the foundation for adult relationships by creating an **internal working model** that acts as a template for the future. This is the **continuity hypothesis** — the belief that similar relationships will occur as an adult.

There are several attachment styles that a child can develop in infancy. Ainsworth (1971), through the 'Strange Situation' methodology, divided these into **secure**, **insecure-avoidant** and **insecure-resistant** styles. Attachment style provides a child with a set of beliefs about themselves and others and the nature of relationships. The continuity hypothesis sees these attachment types as predicting the nature of adult relationships. Therefore someone who is securely attached as a child will have similar relationships throughout life, even with their own subsequent children.

Hazan and Shaver (1987) applied Bowlby's theory to adult relationships, arguing that early attachment patterns affect three areas of adulthood: romantic relationships, care giving and sexuality.

Research

- Simpson et al. (2007) performed a longitudinal study on a group of individuals from childhood into their twenties. Securely attached individuals were more socially competent, developed secure friendships and had positive emotional experiences on a regular basis, supporting the hypothesis.
- Hazan and Shaver (1981) devised a 'love quiz' in a local newspaper, asking readers to describe their feelings and experiences about romantic relationships and their childhood relationships with parents. They found a strong relationship between childhood and adult attachment patterns: for example, individuals with secure attachments had romantic relationships twice as long as insecure types. Insecure-avoidant types doubted the existence of love, feared closeness and found it hard to forgive. Insecure-resistant types were intensely emotional, jealous and untrusting. Secure types believed in love, were trusting and liked being close. These findings support the hypothesis.
- McCarthy (1999) found that women who had insecure-avoidant attachments in childhood did not have successful later romantic relationships, while those with insecure-resistant attachments had poor friendships. Those with secure early attachments had successful romantic relationships and friendships, in line with the hypothesis.
- Schachner and Shaver (2002) found that individuals with an insecure-avoidant attachment type were more accepting of casual sex, lacking intimacy.

Evaluation

- Steele et al. (1998) found a small correlation of 0.17 between having a secure attachment type in childhood and early adulthood, contradicting the continuity hypothesis.
- Attachment types do not appear to be as fixed as first thought. Hamilton (1994) found that securely attached children became insecure as a result of life events.

- Having a childhood insecure attachment style does not seem necessarily to equate with poor-quality adult relationships. Rutter et al. (1999) reported individuals who did not have secure attachments with their parents, but went on to form secure, stable adult relationships.
- Levitt (1991) believes that people have expectations of relationships and although some of these expectations come from previous relationship experiences, not all do, and therefore other factors contribute to the quality of adult relationships.
- The temperament hypothesis sees the quality of adult relationships as not being determined biologically from innate personality. This suggests that attempts to develop better-quality relationships by changing people's attachment styles to more positive ones would not work.
- Although research shows a link between childhood attachment styles and adult relationships, other factors may be contributing, so we cannot establish a causal link.

Interactions with peers

Relationships with peers also have an influence on later adult relationships. Peers become more influential as a child progresses into adolescence, playing a significant role in an individual becoming an independent adult, and helping to develop social skills, including those needed for adult relationships. Peer relationships do not replace adult attachments; they are just another type of attachment.

Peer relationships differ from attachments with adults, as they are **horizontal relationships** between individuals of equal status.

There are two stages to the development of adolescent peer relationships. First, friendship cliques form of small groups of the same sex around 12 years of age. At about age 14, several cliques of both sexes merge together to form groups. From these groups, individuals will form up into romantic couples.

Research

- Meier et al. (2005) found that both type and quality of adolescent relationships relate to type and quality of adult relationships, suggesting a link between the two.
- Hartup (1996) reported that popular children had positive developmental outcomes, while unpopular children did not, which could contribute to the quality of adult relationships.
- Bagwell et al. (1996) found that poor-quality friendships were linked with low self-esteem, while good-quality friendships were linked with high self-esteem, an important factor in having the confidence to build successful adult relationships.
- Hartup (1993) found that popular children with many friends among their peers were more socially able and better at forming relationships.
- Kahn et al. (1985) found that students who had not developed a strong identity, due to poor attachment experiences in infancy, had less success in later relationships. Males were likely not to get married and females were likely to be separated.
- Collins and Van Dulmen (2006) found that experiences in early relationships with both parents and peers influence the quality of young adult romantic

relationships, with both offering opportunities to learn expectations, skills and behaviours affecting relationship quality.

- Connelly and Goldberg (1999) found that the level of intimacy in peer relationships laid the foundations for the degree of intimacy in young adult relationships.

Evaluation

- During adolescence, attachments with parents can change to more of a relationship of equals, similar to relationships with peers, and the development of these attachments can influence adult relationships. Coleman and Hendry (1999) found that children with close parental relationships during adolescence developed the levels of independence required to form successful adult relationships.
- Hartup (1993) thinks it is difficult to calculate the impact of children's peer relationships on adult relationships, as there is a need to take into account who the friends are and the quality of friendships.
- Erikson (1968) believes that the topic area is affected by gender differences. He sees female identity development as dependent on finding a partner, so female identity development comes after intimacy, with the reverse situation for males. This suggests that attachment patterns have different outcomes on adult male and female relationships.
- Wood et al. (2002) believes there is a subtle but important difference between the way in which individuals relate to others, resulting from attachment type, and their relationships, resulting from the interaction of two people's attachment style. Therefore an insecurely attached person could have a secure relationship if they were in a relationship with a securely attached person.
- Attachment theories are somewhat deterministic, as they perceive childhood/peer attachments causing later adult relationships. However, it is likely that other factors play a role, such as the different attachment styles people bring into a relationship.
- Furman (1999) believes that, because peer relationships are on an equal footing, they provide opportunities for cooperation and mutual altruism not present in child–adult relationships, and these qualities are important in forming successful romantic relationships.
- Leaper (1994) believes that girls' and boys' peer groups emphasise different styles of relating, affecting the amount of opportunity for learning skills important for adult relationships. Girls' peer groups emphasise turn taking and mutual decision making and may be more influential, suggesting a gender difference in how peer groups affect adult relationships.

Nature of relationships in different cultures

There are differences between cultures in how relationships are formed and undertaken. In some cultures, such as western ones, partners freely select one another and relationships can be ended if one or both partners wish to. In collectivist cultures, where society as a whole is more important than individual needs, relationships are more permanent and may often be arranged by outside parties. Interestingly, in a worldwide context, arranged marriages are the most common form, with parents being most influential in partner choice, with subsidiary roles played by other family members and friends. Indeed, relationships in these cultures are more of a union between families than between individuals. The idea behind arranged marriages is that young people would choose a partner on the basis of attraction, which is not perceived as a recipe for success. In two-thirds of the world a man, or his family, must pay a dowry for his bride; in return, he gets her labour and childbearing qualities.

Research conducted into relationships tends to be centred on western cultures and may not be applicable to other cultures.

In multicultural societies, like Britain, tensions and conflicts occur, with young people valuing individual choice more, and older generations in some cultures favouring traditional arranged marriages.

Research

- Mwamwenda and Monyooe (1997) found that 87% of Xhosa students in South Africa supported the dowry system, seeing it as a sign of the groom's appreciation for his bride.
- McKenry and Price (1995) reported that, in cultures where females have become more independent and influential, divorce rates have risen considerably, suggesting that the lower divorce rates often seen in non-individualistic cultures are a reflection not of happy marriages, but of male dominance.
- Umadevi et al. (1992) looked at female student preferences for love marriages and arranged marriages in India. Arranged marriages were seen positively as long as the two intended partners consented. However, love marriages were preferred too, so long as there was parental approval, demonstrating the importance of whole family opinions in Indian society.
- Gupta and Singh (1982) looked at 100 Indian marriages of professional, educated couples, 50 of which were arranged marriages and 50 of which were love marriages. Couples were assessed after 1, 5 and 10 years of marriage. In love marriages, loving and liking were initially high, but decreased over time; while in arranged marriages, loving and liking were initially low, but grew and exceeded the level of love marriages after 10 years, suggesting that arranged marriages are more successful over time than love marriages.
- Zaidi and Shuraydi (2002) interviewed Muslim women of Pakistani origin who had been raised in Canada about their attitudes towards arranged marriages. The majority preferred love marriages of their choice, although their elders, especially fathers, were opposed, showing the potential for discord within cultural groups that form minorities within more dominant cultures.

Evaluation

- Simmel (1971) believes that western individualistic cultures have higher divorce rates because individuals are perpetually looking for an ideal partner.
- Xioahe and Whyte (1990) found contradictory evidence suggesting that arranged marriages are superior to love marriages over time. In China, women reported love

marriages more satisfactory than arranged marriages. However, women in China have developed far more freedom and influence in recent years and findings may be an expression of this.

- There is a problem in comparing cultures, as samples are rarely identical, which can invalidate results.
- There are huge variations between cultures in reasons for divorce and these may be a reflection of cultural norms and practices. In Saudi Arabia a woman can only get a divorce if it is a right stated in her marriage contract and she would be loath to do so, as Saudi law states that children remain with the father.
- The dowry system may have a protective value for women, as husbands may be loath to abuse wives they have paid money for. However, it is also used to justify children remaining with the husband's family in the event of divorce, the children having been 'paid for'.
- When researching into cultural differences, cultural bias can be a problem, where researchers interpret observations in terms of their own cultural norms: for example, using measuring tools or questionnaires devised in one particular culture.
- There is an assumption in western culture that inequity in a relationship is a sign of discord, which if unresolved leads to relationship breakdown. However, this is a cultural assumption and is not necessarily true of collectivist societies.
- The concept of totally arranged marriages, where the intended partners have no say in the matter, is a rare occurrence. Most partners in arranged marriages have a right to consent and the majority meet each other at social functions or through a third party.

Aggression

Social psychological approaches to explaining aggression

Specification

- *Social psychological theories of aggression, for example social learning theory, deindividuation*
- *Explanations of institutional aggression*

There is a requirement to have knowledge and understanding of social psychological theories. The two named are examples, meaning they could not be specified in examination questions and any relevant theories would be acceptable.

There is also a requirement to have knowledge of explanations of institutional aggression. Explicit explanations are not specified, so any relevant ones would suffice, such as warfare and terrorism. It would be necessary to be able to describe and evaluate such material.

Social psychological theories of aggression

There are several broad approaches to explaining aggression, such as by reference to biology. However, social psychological theories see aggression arising out of social interactions.

Social learning theory

Although animal aggression tends to be a result of instinctual drives, social learning theory sees human aggression as learned in two ways, both involving **operant conditioning**:

- **direct reinforcement** — a behaviour is reinforced (rewarded), making it likely to be repeated again
- **indirect reinforcement** — an observed behaviour that is reinforced is observed and imitated (**vicarious learning**)

Through social learning, humans learn the value of aggressive behaviour and how and when to imitate specific acts of aggression.

Bandura (1965) outlined four steps of modelling:

- **attention** — attention is paid to attractive, high-status and similar models
- **retention** — observed behaviours need to be memorised
- **reproduction** — imitation only occurs if a person has the skills to reproduce the observed behaviour
- **motivation** — direct and indirect reinforcements (both negative and positive) as well as punishments influence the motivation to imitate

Although a model is necessary for imitation, good levels of **self-efficacy** (situation-specific confidence) are also required.

There are many sources of aggression for social learning to occur through, but media influences form the basis of much research. It has been found that if an observer identifies with the perpetrator of an aggressive act and/or if the act of aggression is more realistic or believable, it is more likely to be imitated. However, if the perpetrator of an aggressive act is punished for their behaviour, it decreases the chances of the behaviour being imitated.

Research

- Bandura et al. (1961, 1963) showed children various scenarios involving aggressive behaviour to a Bobo doll (an inflatable doll that when hit bounces back up). He found that they were likely to behave aggressively and, after being deliberately frustrated, especially imitated specific behaviours they had observed when allowed to play with the doll. Bandura concluded that the chances of aggressive acts being imitated increased if the aggressive model was reinforced, but decreased if the model was punished, suggesting that although observational learning can occur, imitation will only be seen if the behaviour is vicariously reinforced. Aggression was also more likely to be imitated if a child identified with the model (for example, boys were more aggressive if the model was male and to a lesser extent girls were more aggressive if the model was female) or had low self-esteem. Interestingly, a film version of the aggressive behaviour was equally effective.

- Cooper and McKay (1986) found that after children aged between 9 and 10 years of age had played aggressive video games, acts of aggression increased in girls, but not boys, suggesting that such sources of social learning affect genders differently.

Evaluation

- Bandura's studies into aggression are regarded as the cornerstone of social learning theory, but there are several criticisms:
  - **Methodological issues:** Bobo dolls are not real, do not retaliate and are designed to be hit. Some children stated, 'That's the doll we have to hit.'
  - **Personality factors:** Johnston et al. (1977) stated that children who were most aggressive had been rated by their teachers as being more aggressive, suggesting that personality factors are more important than social learning.
  - **Demand characteristics:** the situation was unfamiliar, so the children may have been behaving in the way they thought was expected.
  - **Lack of ecological validity:** the studies were carried out under controlled conditions in a laboratory environment and the artificial set-up may have led to unnatural behaviour.
  - **Measurement of aggression:** Bandura measured the number of aggressive acts, but had no way of distinguishing between real acts of aggression and play-fighting, which may have acted as a confounding variable.
  - **Ethics:** there are doubts about whether informed consent was gained from parents, and as the children were filmed, there are doubts about confidentiality. The main ethical concern, though, is the possibility of distress and the potential long-term consequences for behaviour in encouraging children to be aggressive.
- Huesmann (1988) reported that children use television models to direct their own actions. Observed aggressive acts are stored in memory, where they are strengthened and elaborated through repetition. They are then used to guide behaviour in situations perceived as appropriate, suggesting that media influences are a source of social learning.
- Social learning theory can explain people's levels of aggression varying between situations, by means of them being reinforced to be aggressive, and in varying ways, in different situations. If aggression were biologically determined, we would expect aggression levels to be consistent across situations.
- Social learning theory does not account for biological factors, such as high levels of testosterone. Bandura did acknowledge that biology played a part, but argued that aggressive urges/tendencies are biological, while knowing how and when to be aggressive is learned directly and indirectly.
- Social learning theory does not account for the role of emotional factors in aggressive behaviour.
- Social learning can be argued to be more powerful than biology because there are societies which are non-aggressive, such as the Amish communities in the USA.
- Social learning theory explains individual differences and cultural differences in aggression, as resulting from different learning experiences.

- Bandura's research has been used to explain problem behaviours such as deviancy and eating disorders.
- A practical application of social learning theory is that, if maladaptive behaviours are learned, then alternative adaptive behaviours can be learned in their place.

Deindividuation

Losing their sense of individual identity deindividuates people. Individuals are seen as normally refraining from aggressive acts, because they would be identifiable, but in situations such as crowds, social restraints are less and personal responsibility perceived as lesser, and so aggressive behaviour occurs. The normative view is that deindividuation causes people unquestioningly to follow group norms instead of personal norms and sometimes these group norms lead to aggression.

Zimbardo sees people in crowds as being anonymous, with lessened awareness of individuality and a reduced sense of guilt, or fear of punishment. The bigger the crowd, the more this will be.

Prentice-Dunn and Rogers (1982) believe that individuals normally have awareness of personal moral codes, but being in a crowd diminishes private awareness, so instead they follow the group norms.

Research

- Malmuth and Check (1981) found that nearly a third of US male university students would rape if there were no chance of getting caught.
- Zimbardo (1963) replicated Milgram's electric shock study, but the 'teacher' was either individuated with a name-tag, or deindividuated by wearing a hood. The deindividuated teachers gave more shocks, supporting the idea of deindividuation.
- Diener et al. (1976) found that anonymous 'trick-or-treating' children in the USA stole more money or sweets than non-anonymous children, supporting the notion of deindividuation.
- Silke (2003) found that people who were disguised perpetrated 41% of violent assaults in Northern Ireland. The more severe the assault, the more likely the attacker was to be disguised, suggesting that disguises deindividuate people, reducing guilt and fear of punishment.
- Mann (1981) found that in 48% of incidents of threatened suicide from tower blocks, crowds encouraged the distressed person to jump. This tended to occur when the crowd was large, it was dark and the crowd was some distance away, suggesting that deindividuation was occurring.
- Watson (1973) conducted a cross-cultural study and found that warriors who disguised their appearance — for example, through face paint — tended to be more aggressive, suggesting that deindividuation effects are universal.

Evaluation

- Deindividuation in crowds can lead to increased pro-social behaviour (e.g. religious gatherings).

- The idea that in crowds people lose their personal moral codes is evidently not so, as many people are not negatively affected.
- Deindividuation has been used to explain the phenomenon of football hooliganism. However, Marsh et al. (1978) found that mainly ritualised behaviour occurred, actual violence being rare.
- One practical application arising out of an understanding of deindividuation is the use of closed-circuit television cameras, such as at football matches, which have reduced violence levels.
- Research fails to take into account whether it is the anonymity of the victims being aggressed against, or of the aggressors themselves, which leads to high levels of aggression.
- Recent research suggests that crowds do not cause individuals to lose their sense of identity, but rather individuals adopt localised group norms, creating a common social identity. This explains why Postmes and Spears (1998), when performing a meta-analysis of deindividuation studies, found no consistent pattern of deindividuation affecting individuals' behaviour.

Explanations of institutional aggression

Institutional aggression occurs in two ways:
- **Instrumental aggression** — institutional groups sharing a common identity and aims, such as the police, the army and terrorist gangs, tend to use aggression in a non-emotive manner, as a calculated means of achieving goals.
- **Hostile aggression** — people living in institutions, such as jails, detention centres and care homes, sometimes use aggression emerging from emotional states, like anger and frustration.

Explanations of institutional aggression tend to attribute it to **situational factors**, aggression stemming from factors within the social situation, or **dispositional factors**, where aggression stems from personality factors.

Warfare

Warfare is seen as involving instrumental aggression and occurring through situational factors. However, warfare is not uniquely human. Ant colonies wage war against each other and research into warfare among animal groups has led to evolutionary explanations of warfare being proposed, which see warfare arising from the hunter-killer instinct of carnivores, or from group defence behaviour against predators.

Human warfare can be seen as being destructive, because battle does not take place face to face and therefore humans are divorced from the consequences of their actions. There is also a tendency to dehumanise opponents so that it is easier to aggress against them, and armies deindividuate individuals by the use of uniforms.

Research
- Ardrey (1961) sees human warfare as arising out of evolution of group hunting skills to catch prey animals.

- Kruuk (1972) reported that spotted hyenas regularly wage war over territorial disputes and kill other hyenas hunting prey on their territory, suggesting that warfare arises out of group defence mechanisms.
- Ehrenreich (1997) believes that warfare arose from human groups collectively protecting themselves from attackers and the perception of threatened attack. The 'Indian Wars' in the USA can be interpreted as early colonial settlers wiping out indigenous peoples, because they were perceived as potential attackers.
- Goodall (1986) reported that groups of chimpanzees will wage war on other groups in order to kill and eat them, suggesting that warfare evolved out of group hunting skills.

Terrorism

The main causes of terrorism lie in cultural and subcultural clashes, but can be seen as a form of minority influence, where minority groups seek to affect social change by changing majority views. Their behaviour and beliefs are generally consistent and persistent, leading to gradual changes in public opinion. The Irish Republican Army (IRA) in Northern Ireland had growing popular support among the nationalist community, reflected in the large proportion of the electoral vote achieved by its political wing, Sinn Fein.

Terrorism can be seen to justify aggression by the idea of 'collective responsibility', such as the targeting of random civilians by the London tube bombers in July 2005.

Research

- A Ministry of Defence report (2005) found that in a confidential opinion poll the majority of Iraqis privately, but not publicly, supported the terrorist insurgency, supporting the view that terrorism is a form of minority influence.
- Barak (2004) reported that terrorists are generally people exhibiting suppressed anger who have experienced economic and political marginalisation, suggesting that terrorism has its roots in cultural and subcultural clashes.
- Pyszczynski et al. (1996) got Iranian and US students to focus on their own mortality and found it increased support for extreme violence, such as military interventions and suicide bombings. This suggests a degree of similarity in attitudes and behaviour among both terrorists and their opponents.

Evaluation

- The idea that terrorism results from those experiencing economic and political marginalisation is opposed by the fact that many terrorists, such as the Baader-Meinhof gang and the London tube bombers, were university-educated and from affluent families.
- By their very secretive nature, terrorism and terrorists are difficult to study and present many methodological challenges. Most research 'findings' are based on subjective interpretations and are at risk of extreme researcher bias.
- The idea there is a 'typical' terrorist is rather simplistic. It is more probable that a variety of explanations are needed to explain the vast variety of terrorist groups and

actions, including instrumental and hostile acts, and can be seen to encompass both situational and dispositional factors.

- Merrari (1991) found there was a lack of empirical data and over-reliance on other researchers' results, creating a problem in reaching valid conclusions.
- Schmid and Jongeman (2005) reported an over-reliance on secondary data analysis, problems in defining terrorism and limited data collection and methodologies, giving support to Merrari's findings and suggesting that it is an ongoing, long-term problem.

Prisons

Zimbardo was motivated to undertake his prison study because of reported abuses by both prisoners and guards within the US prison system, which were seen as due to dispositional factors (personality). Volunteers were randomly assigned as guards or prisoners in a mock prison, each having their own distinguishing 'uniform'. Intended to last 2 weeks, the study stopped after 6 days due to the distressed condition of the prisoners and the abusive behaviour of the guards. Zimbardo concluded that behaviour within his mock prison resulted from situational factors, and that brutalising environments produce brutality, suggesting that the potential for such behaviour is within us all.

The real-life abuses of Iraqi prisoners by US troops in Abu Ghraib prison can be explained as the result of similar situational factors and not the 'twisted' disposition of the perpetrators. However, other factors may play a role too: for instance, the abuses occurred after an attempted prison riot in which an officer was hurt, so retaliatory humiliation may have been involved. The abuses were also committed by low-ranking soldiers under no supervision from officers, therefore the abuses committed may have been an attempt to create status and power for those soldiers.

Research

- Haslam and Reicher (2006) reported in the BBC prison study (based on Zimbardo's study) that participants' behaviour could not be explained purely by allotted roles, and that behaviour was better understood in terms of **social identity theory**, which sees behaviour as due to **in-group** and **out-group** reference points (us and them).
- Johnston (1991) found that prison overcrowding leads to increased aggression, due to increased competition for resources and the tendency to adopt violent defensive behaviours, either individually or through the formation of prison gangs with extreme in-group/out-group beliefs. This suggests that situational factors are at play.

Evaluation

- After abuses within institutions become public knowledge, it is common for dispositional explanations to be offered: that is, a few 'bad apples' are to blame. However, this can be perceived as an attempt to find a scapegoat so that the system/institution itself is not seen at fault.
- A lot of research into institutional aggressions concerns the debate over dispositional and situational factors. The best way to view this debate is from an interactionist viewpoint, with personality and situational factors acting upon each other, determining levels and types of aggression.

- There are ethical problems with simulated prison studies, especially the issue of harm. Zimbardo's study was stopped early due to distress and harm, and fully informed consent was not given. However, a cost–benefit analysis could justify such research by the value of the knowledge gained.
- Research permitting a better understanding of aggression in prisons could lead to the formation of useful practical applications, such as prison reform. Zimbardo's research led initially to progressive changes in the treatment of prisoners. However, Zimbardo feels that prison regimes became worse not better, suggesting that his research cannot be justified in cost–benefit terms.

Biological explanations of aggression

Specification

- *Role of neural and hormonal mechanisms in aggression*
- *Role of genetic factors in aggressive behaviour*

The specification here is quite straightforward, centring on two broad biological areas associated with aggressive behaviour. First, biochemical influences via hormonal and neural mechanisms are to be studied and could form the basis of examination questions. The same situation arises with the role of genetic factors — again you will need to be able to describe and evaluate the role that they play in determining levels of aggression.

Neural and hormonal influences

Biological explanations view aggression as having internal physiological causes rather than external social or environmental ones. Research has concentrated on several biological areas, including genetics, hormones, neurotransmitters and brain structures themselves. Biological factors can be perceived as sole causes or as working in conjunction with other factors.

Neurotransmitters and hormones play an important role in many areas of human functioning and the neurotransmitter **serotonin**, along with hormones such as **testosterone**, **adrenaline** and the female hormones **oestrogen** and **progesterone**, have been identified as involved with aggression.

Serotonin

Serotonin has been linked with various bodily effects, such as sleep, but low levels of the neurotransmitter have been especially associated with increased levels of aggression (and high levels with reduced aggression).

Research

- Delville et al. (1997) found that drugs increasing serotonin production lead to reduced levels of aggression, suggesting that high levels of serotonin are linked to increased aggression.

- Linnoila and Virkunen (1992) found a relationship between low levels of serotonin and highly explosive violent behaviours, suggesting that a lack of serotonin is linked to aggression.
- Lidberg et al. (1985) compared serotonin levels of violent criminals with non-violent controls, finding the lowest levels of serotonin among the violent criminals.
- Higley et al. (1996) set up an island colony of 49 2-year-old male rhesus monkeys, regularly measuring their serotonin levels and observing their behaviour. After 4 years, 11 were dead or missing, 46% of which had very low levels of serotonin and regularly exhibited aggressive behaviour. Not one monkey with high levels was dead or missing. Those with high levels stayed in close proximity to each other, indulging in a lot of social grooming. This suggests that serotonin levels affect a continuum of aggressive behaviour from violent through to placid.
- Popova et al. (1991) found that animals selected for domesticity because of reduced aggression levels had lower serotonin levels than wild, more aggressive counterparts.

Evaluation

- Various drugs, approved by clinicians, have been associated with reducing serotonin levels and increasing aggressive behaviour. Penttinen (1995) reports that cholesterol-lowering drugs (e.g. lopid), appetite suppressors (e.g. fenfluramine) and even low-fat diets all produce such effects. Some drugs have been withdrawn because of their potent anti-serotonergic effects.
- The fact that alterations in serotonergic activity induce profound changes in behaviour has been used in positive ways, such as the introduction of Prozac, which works by enhancing serotonin activity and has been used to treat depression.
- Huber et al. (1997) argues that reducing serotonic activity in a wide range of species, from crustaceans to humans, has the same effect of increasing aggression, suggesting an evolutionary link.
- Care must be taken when generalising from animal to humans; many animals do not have the same capacity for cognitive processing and self-awareness. Animal behaviour is often more reliant on biological factors.
- Evidence linking low levels of serotonin and aggression is strong and widespread, but is only correlational and does not indicate causality.

Testosterone

Testosterone is a male hormone (though found in low levels in females), the presence of which is associated with aggression. Castration, which lowers testosterone levels, generally reduces aggression. Studies where testosterone has been given to females indicate heightened levels of aggression. However, effects are not universal and some research has not found links between heightened testosterone levels and aggression in humans.

Testosterone modulates levels of various neurotransmitters that mediate effects on aggression. There seems to be a critical period early in life, where exposure to testosterone is essential to elicit aggression in adulthood. It is thought testosterone helps sensitise an androgen responsive system (responsive to male hormones).

Research

- Connor and Levine (1969) found that rats castrated after birth had reduced aggression when tested as adults, suggesting a link between aggression and testosterone. However, rats castrated after puberty could be made aggressive with injections of testosterone, while the early castrated rats could not, suggesting a developmental factor.
- Edwards (1968) found that injections of testosterone in neonate female mice made them act like males with increased aggression, when given testosterone as adults. However, control females only given testosterone as adults did not react in this way, suggesting that testosterone masculinises androgen-sensitive neural circuits underlying aggression in the brain.
- Nelson (1995) reviewed research into hormonal affects on aggression in humans and reported a link between high levels of testosterone and violent behaviour in males and females in prisons, although levels of the hormone were tested subsequent to aggressive behaviour, invalidating findings somewhat.
- Bermond et al. (1982) found that testosterone only affects certain types of aggression in animals, such as intermale aggression as a defence response to intruders, while predatory aggression is not affected.
- Albert et al. (1993) found that raising testosterone levels in female rats elicited aggressive behaviour in the presence of certain environmental events, such as competition, suggesting that testosterone alone is not responsible for aggression, but requires certain environmental triggers.
- Higley et al. (1996) reported that individuals with raised levels of testosterone exhibit signs of aggression, but rarely commit aggressive acts, suggesting that social and cognitive factors play a mediating role.

Evaluation

- Although strongly linked to aggression, testosterone is only one factor and the effects of previous experience and environmental stimuli have been found to correlate more strongly.
- Results from human studies are often subjective, relying on questionnaires and observations.
- One problem with using animal studies to research into testosterone affects is that certain brain structures are involved with different types of aggression in different species. For example, the cingulated gyrus is linked to fear-induced aggression in monkeys, but to irritability in cats and dogs, creating problems in trying to generalise to humans.
- Castration research usually indicates reduced aggression. However, castration disrupts other hormone systems as well as testosterone and these may be playing a part too.
- There are ethical issues in eliciting aggressive behaviour among animals, but justification may centre on cost–benefit analyses, where the value of the knowledge gained could be seen to outweigh the distress caused.

- Albert et al. (1993) reports that there is a lot of research that does not show a significant relationship between aggression and testosterone. This may be because testosterone is only linked to certain types of aggression, or other factors are involved, or there were methodological issues.

Role of genetic factors in aggressive behaviour

Research has indicated that various genes may be associated with aggression. The MAOA gene, which helps to eliminate excess amounts of some neurotransmitters, such as serotonin and dopamine, has been implicated as well as the sex chromosome gene Sry.

Research

- Moffitt et al. (1992) performed a longitudinal study on 442 New Zealand males from birth to age 26, recording which participants, as children, suffered abuse and also what level of activity of the MAOA gene participants had. It was found that those who had suffered abuse and had the low activity version of the gene were nine times more likely to indulge in anti-social behaviour, including aggression. Participants who had been abused, but carried the high activity version of the gene, were no more likely to be anti-social than those not suffering abuse, suggesting that the MAOA gene is involved in aggressive behaviour.
- Brunner et al. (1993) studied a large Dutch family where all the males had a mutant form of the MAOA gene. All had borderline retardation and reacted aggressively when angry, fearful or frustrated, suggesting that abnormal MAOA activity is associated with aggression.
- Cases et al. (1995) studied mice genetically engineered to lack MAOA. The mice had a dramatically altered serotonin metabolism and severe behavioural alterations. When adult they showed enhanced aggression and were aggressive during mating, giving support to human studies suggesting that aggression is a direct result of MAOA deficiency, rather than other genetic influences or psychosocial factors.
- Rissman et al. (2006) performed research into Sry, a gene that leads to the development of testes and high androgen levels in males. Both male and female mice with and without the gene were tested, with the Sry gene being associated with high levels of aggression. The research suggests that genes and hormones interact with each other, and although sex chromosome genes are not the whole story, they seem to play a big part.
- New et al. (2003) found that acts of impulsive aggression, such as domestic and work-based attacks, have a genetic component related to the serotonergic system. Aggressive patients with personality disorders were found to have a G-allele variant of a serotonin gene, HTR1B, suggesting that many genes may be involved in aggressive behaviour.
- Deneris et al. (2003) removed the PET 1 gene, linked to production of serotonin, from mice, finding that these normally placid creatures became anxious and aggressive, especially to intruders entering their territory. The behaviour of the

mice was similar to human personality disorders characterised by anxiety and violence, suggesting that over-aggressive people may be missing this gene.

Evaluation

- One-third of males carry the low-level activity version of the MAOA gene, suggesting that it bestows adaptive advantages. Also associated with risk taking, it may have beneficial qualities in certain occupations, such as stockbroking.
- It may be possible to devise drug treatments for people with the low-level activity version of the MAOA gene to help control aggressive urges. However, Tyrer et al. (2008) reports that drug treatments have so far not had success in controlling aggression levels, suggesting that other non-biological factors may be involved.
- There is a mistaken belief that genes can determine aggressive behaviour on their own and this has formed the 'natural born killer' defence in murder trials. Stephen Mobley, accused of murder in the USA in 1991, pleaded not-guilty, claiming that he had a gene predisposing him to violence. His argument was that several members of his family were aggressive and he had inherited the tendency. However, he was executed.
- New (2003), on the basis of research into genes related to the serotonic system, has hypothesised that multiple genes and environmental factors contribute to an individual's degree of susceptibility to impulsive aggression. This suggests that an interactionist explanation utilising biological and environmental factors works best.
- There is a problem generalising from animal studies to humans, as similar results are not always found. For example, mice lacking the HTR1B gene, associated with serotonin production, had elevated levels of aggression, but the same was not found in humans.

Brain structures

Various brain structures have been associated with aggressive behaviour, with the cortex seen as playing an inhibiting role over the sub-cortical limbic system, preventing aggression. The amygdala (part of the limbic system) has been seen as having an important role.

Research into violent criminals has linked abnormalities in their limbic systems to murderous behaviour, and tumours within the limbic system have been seen to elicit aggression.

Research

- Bard (1929) removed the cortex from cats' brains, eliciting anger without emotional content ('sham' rage), suggesting that the cortex plays an inhibitory role over the limbic system, regulating aggression.
- Egger and Flynn (1967) found that stimulating one area of the amygdala produced an inhibitory effect, but lesioning it led to increased aggression. This suggests that the amygdala plays a key role in regulating aggression and the cortex has inhibitory control over it (the amygdala being sub-cortical).
- Raine et al. (1997) scanned murderers' brains, finding that they were more at risk of having abnormalities in their limbic systems. This suggests that the limbic system is involved in the control of aggression.

- Sumer et al. (2007) reported on a case study of a patient with a tumour in their limbic system exhibiting increased aggressive hyperactivity. When the tumour was treated, the patient returned to normal.

Evaluation

- Stimulating different areas of the amygdala can inhibit or increase aggression, suggesting a complex relationship involving different forms of aggression. For example, Hernandez-Peon et al. (1967) elicited affective rage and flight, but not predatory attack, by applying the neurotransmitter acetylcholine to the amygdala.
- The relationship of the limbic system to aggression is a complex one, with different areas being implicated in different forms of aggression.
- The research linking brain abnormalities to violent crime is only correlational, and although having such abnormalities may increase an individual's vulnerability to being aggressive, other factors may be involved.

General evaluation of biological explanations

- Any explanation that sees biology alone as determining aggression is a deterministic one, perceiving no role for free will. Although this may be true for lower animals, human behaviour also involves cognitive input.
- Biological explanations can be seen as reductionist, reducing explanations down to the single factor of biology. This may be rather simplistic, as research suggests that aggressive behaviour involves a complex interaction of factors.
- Biological explanations cannot explain cultural differences in types and levels of aggression, although they can suggest explanations for differences between men and women.
- Biological explanations can lead to better understanding, allowing development of positive practical applications, such as drug treatments for aggressive behaviour.

Aggression as an adaptive response

Specification

- *Evolutionary explanations of human aggression, including infidelity and jealousy*
- *Explanations of group display in humans, for example sports events and lynch mobs*

The specification centres on evolutionary explanations of human aggression. Infidelity and jealousy are explicitly named, so there is a requirement to have sufficient knowledge of them both to describe and to evaluate them, as they may feature directly in examination questions.

There is also a requirement to focus on group display in humans, but sports events and lynch mobs are just examples and so would not feature explicitly in exam questions; any relevant material would be creditworthy.

Evolutionary explanations of human aggression

Evolutionary explanations see aggression as having an adaptive advantage and therefore, through natural selection, becoming widespread.

Lorenz (1966) saw aggression as an instinct that had evolved in all species, bestowing a survival value in allowing animals to control access to important resources, such as gaining and securing territories, mating and food. Lorenz believed that environmental cues, such as the presence of an intruder, activated the release of **action-specific energy**, which at certain levels triggers aggressive behaviour. Lorenz believed that a lot of aggressive behaviour had evolved into rituals, such as males' dominance fights, which protect animals from incurring serious harm, one animal generally backing down.

More recent evolutionary theories focus upon aggression as a means of solving adaptive problems covering a wide range of behaviours, between and within species.

- **Between species.** Aggression here generally occurs through predation, with prey animals using aggression to defend themselves and their offspring. Prey animals have also evolved the ability to gauge the strength of predators and know when they are outnumbered. Therefore the 'flight or fight' response is dependent upon this calculation. Altruistic alarm calls to the presence of predators have also evolved as a prey defence mechanism.
- **Within a species.** Aggression between members of the same species can serve several purposes, the most common being to establish dominance hierarchies. As it is generally males who compete for access to females, it is males who are more aggressive, against one another and even against other males' offspring.

Evolutionary explanations see levels and types of aggression occurring due to **evolutionary stable strategies** (ESS), where a balance of different aggression strategies used by different individuals within a population allows population numbers to remain stable. If all members were equally violent, they would wipe each other out.

The evolution of aggression in humans occurred in the Pleistocene era (EEA) as a response to the environmental demands. Therefore aggressive behaviour is a legacy of our ancestral past.

Jealousy

Jealousy occurs through fear of losing affection or status, is characterised by feelings of resentfulness, bitterness and envy, and can motivate aggressive behaviours.

- **Male–male rivalries.** Males compete to be chosen as partners by females, often in ritualised ways.
- **Female–female rivalries.** Females compete to be attractive and will criticise other females' appearances, as males value attractiveness as a sign of fertility.
- **Sibling rivalry.** Siblings compete for parental resources to maximise adaptive fitness.

Research

- Daly and Wilson (1988) reported that male–male rivalries involving aggression are found among young males in nearly all cultures, suggesting that the behaviour is universal as a result of evolution.
- Buss and Dedden (1990) found that females criticise the appearance and sexual promiscuity of other females, suggesting they are reducing their potential rivals' attractiveness, and raising their own by doing so, in line with evolutionary theory.
- McNerney and Usner (2001) found that, although parents encourage siblings to share, they often meet resistance. This suggests that parents are trying to maximise resources in line with evolutionary theory.
- Barrett et al. (2002) used Viking blood feuds to show how having aggressive 'berserker' members in a family group benefited survival, suggesting that aggression has been naturally selected.

Evaluation

- Not all researchers see predation as aggressive behaviour. It lacks an emotional component and the areas activated in the hypothalamus during predation are similar to those associated with hunger rather than aggression.
- Harris (2004) believes that sibling jealousy evolved to maximise parental resources and that this explains the origins of jealousy better than infidelity.
- The tendency towards violent acts, even murder, over apparently minor territorial disputes between neighbours, is explained in evolutionary terms as territorial defensive behaviour involving male–male rivalries.

Infidelity

Infidelity is seen in terms of sexual unfaithfulness, but can also encompass emotional loyalty and fidelity. Evolutionary theory sees cues initiating sexual jealousy as being weighted differently between males and females. Because males are not sure of paternity, they fear sexual infidelity in their partner, and any sign of it (e.g. smiling at other men), especially if they are attractive, resource rich or high status, triggers sexual jealousy and initiates aggressive behaviours to ensure sole sexual access to the male's partner. A male's sexual infidelity does not threaten females' certainty of maternity, but she does fear male emotional involvement with other women in case he spends time and resources on them, which she would prefer to be spent on her and her children. Therefore deprivation of emotional support triggers jealousy in women. Jealousy can also be initiated by the presence of younger, more attractive women.

Research

- Buunk et al. (1996) found that women become jealous when partners become interested in other women, resulting in loss of commitment to the woman and her children, which is essential to survival.
- Looy (2001) found that jealousy in women is triggered by the presence of younger, more attractive women, in line with evolutionary predictions.
- DeSteno and Salovey (1996) found that emotional infidelity is doubly distressing as it also implies sexual infidelity.

- Goetz et al. (2008) looked at men's violence against intimate partners and found that violence functions to punish and deter female sexual infidelity and acts as an anti-cuckoldry tactic, its frequency being related to suspicions of sexual infidelity.
- Daly et al. (1982) found that men are violent when partners are sexually unfaithful, supporting evolutionary theory.
- Buss et al. (1992) measured stress levels in US students, finding that males had higher levels when viewing pictures of sexual infidelity, while females had higher levels when viewing pictures of emotional infidelity. This suggests that different environmental cues trigger aggression in males and females.

Evaluation

- The evolutionary perspective offers an explanation of how aggressive behaviour due to suspicions of infidelity may arise via natural selection.
- Harris (2003) found Buss et al.'s findings about males being more stressed by sexual infidelity and females by emotional infidelity to be true of imagined scenarios, but in real instances both males and females felt threatened by emotional infidelity. The results from the imagined scenarios might be explained as males being aroused by images of sexual infidelity rather than feeling threatened.
- Cultural differences in murder rates of wives by husbands and in the degree of anxiety felt in response to sexual infidelity by males, suggest that other factors besides evolutionarily determined ones play a part.
- Some critics feel that evolutionary explanations justify violence by men against women as natural and inevitable.
- Predictions can be made from evolutionary theory about expected behaviour, and generally research has supported these predictions, supporting evolutionary explanations.

Explanations of group display in humans

Group displays are ritualised displays of aggression, serving the functions of determining dominance hierarchies in relation to ownership of territory and intimidation of other groups. Sports events include a number of features demonstrating group displays; indeed, many aspects of sporting competitions serve as a vehicle for group display, both on and off the pitch.

Lynch mobs are another example of group display, where temporary groups of people are involved in the common pursuit or undertaking of some violent act.

Sports events

- **War dances/supporter displays.** These are performed before battle by warriors, to intimidate the enemy and motivate. They are incorporated into sporting occasions to serve the same purpose, such as the New Zealand haka, performed before kick-off by the national rugby league and union sides. Other sports have developed specialist dance troupes to rouse the crowd's emotional support and intimidate the opposition, such as the Dallas Cowboys' cheerleaders. Football clubs use mascots to the same end and there have been examples of mascots coming to blows. Elements of war dances have been incorporated into supporter displays, as in the wearing of club colours, face painting and club anthems, such as Liverpool supporters singing, 'you'll never walk alone'. Such displays are motivating and increase social identity

- **Territorial behaviour.** Group displays serve to mark out and defend territories: for example, football team supporters congregating in traditional areas and violently resisting attempts by opposition fans to occupy the same territory. At a game in the 1970s, visiting fans arrived early to occupy an area traditionally held by Stoke City fans, and in the ensuing, predictable brawl, crush barriers collapsed and several fans were killed. In most sports there is a noticeable home advantage, due in some part to territorial behaviour.
- **Ritual behaviour.** A lot of aggressive behaviour between rival sports fans is ritualistic, where a lot of posturing and verbal abuse occurs, but little actual violence, suggesting that it is a symbolic show of strength designed to limit injuries. This serves to bond and motivate and is linked to territorial behaviour (see above).

Research

- BBC News (2001) reported that Cyril the Swan, club mascot of Swansea City, was fined £1,000 for fighting with Zampa the Lion, mascot of Millwall, pulling off his head and drop-kicking it into the crowd, sparking a mini-riot.
- Sua Peter (2007) reported that the Siva Tua, the traditional war dance performed by Samoan rugby players before a match, was to be upgraded to a more aggressive, intimidatory style, reflecting the islanders' warrior traditions.
- Shwarz and Barkey (1977) believe that sports teams win more games at home due to the social support of the home supporters, suggesting that territorial group displays may be a factor.
- Morris (1981) conducted a non-participant observation of Oxford United fans, home and away, finding behaviour extremely territorial and ritualised. This suggests that such group displays serve a social purpose.
- Maynard-Smith and Parker (1976) found that territorial ownership by animals was an evolutionarily stable strategy, suggesting that it may have evolved.
- Marsh (1982) conducted observations of football fans, finding that most aggression was verbal, symbolic, non-serious and harmless, serving to reduce levels of aggression. This suggests that most group displays act as a form of catharsis, allowing safe release of negative emotions.
- End (2005) found that the environment of sports events encourages aggressive group displays, suggesting that they are a social construction.
- Grieve (2005) believes that identification with sports teams is psychologically important in an increasingly transient and insular society, suggesting that group displays allow individuals to feel a sense of social identity.

Evaluation

- The universal nature of war dances cross-culturally in sport suggests that the behaviour may have an evolutionary component related to ritualised aggression.

- Many sports teams' war dances are artificial and constructed for commercial purposes, and do not reflect traditional practices.
- Marsh (1982) believes that if ritual aggressive practices between fans were curtailed, violence rates would increase.
- Dunning et al. (1988) argue that, far from being ritualised and harmless, a lot of aggression at sports matches is violent, resulting in many deaths.
- Although group displays may be a factor in aggression levels related to sports events, there are other explanatory factors too, including biological and cognitive ones.
- Group displays at sports events can be explained as a socialisation process, serving to emphasise social identity.
- There are methodological challenges in studying group displays at sporting occasions. Berk (1974) notes that crowd events happen quickly, often without warning, often simultaneously, over a wide area, and involve processes leaving few traces. It would also be dangerous and difficult to interview people during such displays and subsequent accounts tend to be unreliable.
- Guttman (1986) believes that no single explanation can cover the behaviour of sports crowds' behaviour, as they differ so widely.

Lynch mobs

Lynch mobs are associated with unlawful group action, resulting in death. Between 1852 and 1955 there were over 3,000 recorded lynchings of black people by white mobs in the USA. Such mob behaviour was described by Le Bon's (1903) **contagion theory** as being irrational, unthinking, characterised by anger and occurring as a result of **deindividuation**, where an individual loses their sense of identity. He fundamentally saw mob behaviour as a *social contagion* spreading quickly, turning crowds into collections of people indulging in unconscious acts of violence.

Convergence theory sees crowd behaviour as behaviour constructed by particular individuals within a crowd. Crowd behaviour is seen as a convergence of similar-minded people, who wish to act in that way. Lynch mob behaviour is seen as occurring because of popular, deep-seated hatred of those being lynched.

Turner and Killian's (1957) **emergent-norm theory** sees crowd behaviour as the desired, collective actions of people, directed by norms emerging at the time, with different individuals taking on different roles. This means that crowds can be unpredictable, as norms tend to arise spontaneously.

Research

- Waddington et al. (1987) compared violent and non-violent miners' rallies, finding that police actions led to violence and violence was dependent on the social context, not the characteristics of the crowd.
- Marsh et al. (1978) observed football fans, finding behaviour to be structured and norm-related, suggesting that crowds share a common purpose.
- Aguirre et al. (1998) used self-reports to investigate crowd behaviour during the explosion at the World Trade Center in 1993. They found behaviour to be consistent with emergent norm theory, lending support to the theory.

- Turner and Killian (1957) reported that not all members of crowds behave in the same way, casting doubt on to Le Bon's idea of social contagion.
- Mullen (2007) analysed archive material to investigate lynch mob behaviour. Group numbers were noted, plus the severity of atrocities committed and the duration of the actions. Size of lynch mob was positively correlated with the severity of the atrocities.
- Berk (1974) found that convergence theory could not explain all crowd behaviour, suggesting that it lacked theoretical structure.

Evaluation

- The media tend to focus on riotous crowd behaviour and ignore peaceful crowds, giving the wrongful impression of crowds as contagious mobs.
- Research support for emergent-norm theory suggests that the theory explains the difference between relatively unpredictable collective behaviour and more predictable institutionalised behaviour.
- Brown and Lewis (1998) applied existing theories to anti-Vietnam riotous crowds of the 1970's, concluding that no one theory could explain all behaviour, demonstrating the non-universal nature of such crowds.
- Nye (1975) reported that Le Bon's contagion theory was popular and was utilised by important figures such as Mussolini and Hitler due to its emphasis on racial factors.
- Bendersky (2007) used military records, showing that Le Bon's contagion theory had influence on US military thinking and practice throughout the Second World War.
- Turner and Killian (1957) criticised Le Bon's use of the terms 'irrational' and 'emotional', believing that 'unpredictable' and 'spontaneous' were better.
- Wright (1978) criticised emergent-norm theory as being incomplete and having limited practical applications.
- The theoretical nature of explanations of lynch mobs reflects the philosophical nature of the debate. Such actions can be seen either as unconscious deterministic ones, or as motivated and regulated by one's own free will.

Cognition and development
Development of thinking

Specification

- *Theories of cognitive development, including Piaget, Vygotsky and Bruner*
- *Applications of these theories to education*

The specification divides into two clear, separate, but connected parts. As Piaget's, Vygotsky's and Bruner's theories are explicitly named, they could figure in the wording of examination questions, so you should be able to describe and evaluate each of these theories fully. Second, a similar approach must be taken to how these theories can be applied specifically to education, and again you should be able to describe and evaluate the applications.

Theories of cognitive development

Theories of cognitive development attempt to explain the growth of mental abilities. Some theories see thought processes as undergoing qualitative changes as children age, with biological processes directing these changes. Other theorists believe that learning experiences are the major influence. The relative influence of innate and environmental factors is a key issue.

Piaget

Piaget produced a theory of biological maturation seeing *qualitative* differences between adult and child thinking, with a set sequence of developmental stages and children only progressing when biologically ready (see Table 4).

An infant is born with basic reflexes and innate **schema** (ways of understanding the world). If new experiences fit the existing schema, they are *assimilated*. If they do not, they create **disequilibrium** (an unpleasant state of imbalance) and **accommodation** occurs where the existing schema changes to fit in new experiences.

Assimilation and accommodation are **invariant processes** (remain the same).

Schemas and **operations** (internally consistent, logical mental rules) are **variant structures** (change as an individual develops).

A desire for **equilibrium** (pleasant state of balance) drives the process of development, with new experiences pushing children into disequilibrium.

Table 4 Piaget's stages of development

Stage of development	Description
Sensorimotor stage (0–2 years)	New schemas arise from matching sensory to motor experiences. Objective permanence occurs.
Pre-operational stage (2–7 years)	Internal images, symbols and language develop. Children are influenced by how things seem, not logic.
Concrete operational stage (7–11 years)	Development of conservation (use of logical rules), but only if situations are concrete, not abstract. Decline of egocentrism.
Formal operational stage (11+ years)	Abstract manipulation of ideas (concepts without physical presence). Not achieved by all.

Research

- Piaget (1954) found that 3–4-month-old babies do not look for items out of view, suggesting that they have no object permanence. Bower and Wishart (1972) disagree, stating that 1-month-old babies show surprise when items disappear.

- Piaget and Inhelder (1956) suggested that children under 7 years of age were *egocentric*, as they chose the view they could see of a model of mountains, rather than the doll's view asked for. Donaldson (1978) said that mountains are unfamiliar to children and 4-year-olds could see from another's point of view using a scenario based on familiar hide-and-seek.
- Piaget (1956) found that if one of two equally spaced lines of counters are stretched out, children under 7 years of age believe the amounts are now different. McGarrigle and Donaldson (1974) found that if 'naughty teddy' moved them, children agree they are still the same, suggesting that children believe from experience a different answer is required when an adult moves things.

Evaluation

- Due to poor methodology, Piaget underestimated what children could do.
- Piaget's theory functioned as a starting point for subsequent theories and research.
- The idea of separate stages is misleading: individuals can straddle more than one stage at once.
- Piaget believed that the rate of development could not be accelerated, but Meadows (1988) found that direct tuition speeded up development.
- Cross-cultural evidence suggests that the sequence of development is invariant and universal (except for formal operations), suggesting that it is a biological process of maturation.
- Piaget over-emphasised cognitive aspects at the expense of emotional and social factors.
- Piaget saw language ability as reflecting the level of cognitive development, while Bruner sees language development as preceding cognitive development.

Vygotsky

Vygotsky saw the input of cultural knowledge as central to development. He believed that knowledge and thinking were socially constructed by children interacting with people from their culture.

Culture is seen as changing **elementary mental functions** (innate capacities such as attention) into **higher mental functions**, such as comprehension of language. Therefore, culture teaches children what and how to think and there are several ways in which culture can influence cognitive development.

The **zone of proximal development** (ZPD) is the distance between current and potential ability. Cultural influences and experts push children through the ZPD and on to tasks beyond their current ability.

Scaffolding involves being given clues rather than answers. At first, learning involves shared social activities, but eventually individuals can self-scaffold and learning becomes an individual, self-regulated activity.

Semiotics help cognitive development through the use of language and other cultural symbols. These act as a medium for knowledge to be transmitted, turning elementary mental functions into higher ones.

At first, children use **pre-intellectual language** for social and emotional purposes and **pre-linguistic thinking** occurs without language. From 2 years of age, language and thought combine.

- **Social speech (0–3 years)** — pre-intellectual language.
- **Egocentric speech (3–7 years)** — self-talk/thinking aloud.
- **Inner speech (7+ years)** — self-talk becomes silent and internal, language is used for social communications.

From research Vygotsky proposed four stages of **concept formation** (Table 5).

Table 5 Vygotsky's five stages of concept formation

Stages of concept formation	Description
Vague syncretic	Trial and error formation without comprehension. Similar to Piaget's pre-operational stage.
Complex	Use of some strategies, but not very systematic.
Potential concept	More systematic with one attribute being focused on at a time (e.g. weight).
Mature concept	Several attributes can be dealt with systematically (e.g. weight and colour). Similar to Piaget's formal operations.

Research

- Vygotsky (1934) gave children blocks with nonsense symbols on them and they had to work out what the symbols meant. Four different approaches were observed, from which he devised stages of concept formation.
- Gredler (1992) reported that in New Guinea the symbolic use of fingers and arms when counting limited cognitive development, supporting the idea of cultural influence.
- McNaughton and Leyland (1990) observed mothers giving increasingly explicit help to children assembling progressively harder jigsaws, supporting the idea of scaffolding and suggesting sensitivity to a child's ZPD.
- Wertsch et al. (1980) found that the amount of time children under 5 years of age spent looking at their mothers when assembling jigsaws decreased with age, supporting the idea of increased self-regulation.
- Berk (1994) found that children talked to themselves more when doing difficult tasks, supporting the idea of egocentric speech. This decreased with age.

Evaluation

- There is a lack of research support for the theory, but as it focuses on processes rather than outcomes, it is harder to test.
- The theory overemphasises the role of social factors at the expense of biological and individual ones. Learning would be faster if development depended only on social factors.
- The theory is more suited to collectivist cultures with more stress on social learning.
- There are strong central similarities between Piaget's and Vygotsky's theories and an integration of the two may be feasible and instructive.

Bruner

Bruner was influenced by Piaget, but was more interested in the mental representation of knowledge than cognitive development, believing that children have an innate biological organisation aiding their understanding of the world. This develops in increasingly complex ways, allowing more complex thinking, with learning occurring accidentally as children interact with their world. Learning is seen as taking place by organising information into categories, categories being further organised into hierarchies, with the more general at the top.

Bruner proposed three **modes of representation** or ways of cognitively representing the world (see Table 6).

Table 6 Bruner's modes of representation

Mode of representation	Description
Enactive (early childhood)	Knowledge is based on actions (muscle memories). Similar to the sensorimotor stage.
Iconic (middle childhood)	Images of things and events experienced are built up.
Symbolic (adolescence)	Development of mental symbols (language) and logical thought.

The transition from *iconic* to *symbolic* mode at around 6 or 7 years of age is crucial. Bruner sees cognitive development as dependent on language development and a child's level of thinking as determined by its level of language.

Research

- Bruner and Kenney (1966) showed the transition from the iconic to the symbolic mode by means of **transposition**. Most 5- and 7-year-olds could memorise a matrix of nine glasses and reproduce them (iconic mode). When asked to transpose the glasses (given a reverse image of the glasses) no 5-year-olds, but 79% of 7-year-olds could do it (symbolic mode), suggesting that only older children could think using images and language.
- Sonstroem (1966) used 5- and 6-year-olds who could not conserve, finding that children who had to reshape a plasticine ball and describe the new shape, could conserve when tested again because they had been encouraged to use language skills (symbolic mode) in combination with motor skills (enactive mode). The appearance of the plasticine (iconic mode) ceases to dominate, so they can conserve, supporting Bruner's theory.

Evaluation

- Research on deaf children by Furth (1966) showed the development of normal thinking without using language. This suggests that language development is not necessary for cognitive development, supporting Piaget not Bruner.
- Hatwell (1966) found that blind children with normal language skills, who are impaired in sensorimotor experiences, develop operational thought more slowly.

This suggests that language is not necessary for cognitive development, supporting Piaget not Bruner.
- Bruner's theory stresses the role of biology and experience, language and social factors.
- Bruner agreed with Vygotsky that tutoring could develop potential.

Application of theories to education

All three theories have educational applications, and a wealth of research has been conducted to assess their effectiveness in actual classroom environments.

Piaget
- **Concept of readiness.** A child is seen as not capable of learning something until ready to do so, limiting what can be learned at a certain time. Teachers should teach age related material in the order of development.
- **Discovery learning.** Learning is child-centred by the child interacting with its environment and constructing knowledge itself. The teacher creates disequilibrium, making children accommodate new experiences and develop their schemas.
- **Role of the teacher.** Each child's stage of development is assessed and suitable tasks given to challenge the child, pushing it into disequilibrium. Relevant materials are provided at different ages. Opportunities for small group learning are given as learning occurs from conflicting views and peer interactions have social as well as cognitive value.

Research
- Modgil et al. (1983) found that discovery learning leads to poor reading and writing skills in children who need assistance.
- Danner and Day (1977) found that coaching 10- and 13-year-olds had no effect, supporting Piaget's concept of readiness. However, it did assist 17-year-olds, suggesting that tuition helps at a later stage of development.
- Meadows (1988) found that direct tuition speeded up development, contradicting other findings. This suggests that researcher bias plays a part in this contentious area.
- Driscoll (1994) found that a number of instructional strategies have been derived from Piaget's theory, including provision of a supportive environment, utilising social interactions and peer teaching, as well as guiding children to see errors and inconsistencies in their thinking.

Evaluation
- Piaget never intended his theory as an educational tool; it was others who put it to this use.
- Piaget's theory had a big influence on education. The Plowden report recommended that primary education move from being teacher-led to child-centred.
- The Piagetian idea of manipulating concrete materials when learning about abstract principles formed the basis of the Nuffield secondary science project and the Montessori approach to teaching, with some reported success.

- Walkerdine (1984) believes that educationalists used Piaget's theory as a convenient vehicle to justify changes they wished to make.
- Piaget's research sample was small and consisted of children of well-educated professionals, suggesting that his findings may not suit the educational needs of other types of children.

Vygotsky

- **Cooperative and collaborative thinking.** Knowledge is socially constructed by learners working collectively on a common task, where all individuals depend upon and are accountable to each other. This helps individuals to then work better on their own.
- **Peer tutoring.** Allowing peers to be tutors can create a beneficial learning experience for the tutor.
- **Expert tutoring.** This is seen as an effective teaching tool if the boundaries of a child's ZPD are taken into account.
- **Scaffolding.** Experienced people can assist development, providing general and specific tutoring, and enabling individuals to achieve more. Eventually, scaffolding becomes self-instruction.

Research

- Wood et al. (1976) observed mothers and children working together, finding that scaffolding worked best when general encouragement was given if a child was working well, and when specific instructions were given if the learner was struggling.
- Bennett and Dunne (1991) found that children who worked in cooperative groups displayed more logical thinking and were less competitive or interested in status, supporting Vygotsky's concept of collaborative thinking.
- Cloward (1967) found that peer tutoring had more learning benefit for the tutor than the designated learner, demonstrating the benefit of the method.
- Gokhale (1995) tested students on critical thinking, finding that those undertaking collaborative learning outscored those who studied alone, supporting the idea of cooperative learning.

Evaluation

- Vygotsky's teaching methods are dependent on teachers being expert in recognising the boundaries of ZPDs as well knowing how and when to give tuition. This may be unrealistic.
- Some children do not benefit from collaborative thinking and learn best alone, suggesting that individual differences are a factor.
- Learning via cooperative groups needs careful monitoring or some individuals will dominate and others coast.
- Vygotsky's approach may work less well in individualistic settings, where the emphasis is on competitiveness and being autonomous.

Bruner

- **Discovery learning.** As with Piaget, emphasis is placed on the active role of the learner, but teachers have a more central role, not only offering direct tuition, but also providing strategies for self-learning, and identifying specific cognitive abilities required for a task and any consistent errors that learners make. Pupils learn to organise information by themselves, integrating it into existing hierarchical structures or adapting it into new hierarchies. Learning is seen not just as accumulating information, but as creating structure for the information.

- **Scaffolding.** Adult and peer tutors are used to give tuition, which enables pupils to achieve more.
- **Spiral curriculum.** Concepts are mastered by revisiting them at different ages, redeveloping them with increased complexity when more mature modes of thought permit.
- **Materials and activities.** Children are provided with study materials, activities and tools matching and cultivating cognitive abilities. There is a transition as children progress through different modes of thought, from using concrete, to pictorial and then symbolic activities, creating more effective learning. The use of appropriate and stimulating materials and activities creates motivation for learning to occur.

Research

- Gray and Tall (1994) showed how children can use symbolism as a concept and a process to learn algebra and arithmetic successfully, suggesting that Bruner's theory can be applied to the teaching of mathematics.
- Leat (1998) found Bruner's methods incorporated successfully into the teaching of geography.
- Smith (2002) reported that Bruner's idea of intuitive and analytical thinking to solve problems was introduced into schools and colleges with a degree of success.

Evaluation

- Bruner's theory has educational uses beyond schools and colleges. Xerox PARC researchers used it to create graphical use interfaces, addressing the enactive, iconic and symbolic ways in which users understand and manipulate the world around them.
- Bruner was a central member in founding and teaching the colloquium on the theory of legal practice, involving the study of how law is understood.
- The idea of a spiral curriculum as well as Bruner's idea of structured learning has met with a lot of praise.
- Bruner's work greatly influenced Gardner's construction of his theory of multiple intelligences.
- Bruner successfully incorporated discovery learning with scaffolding.

Development of moral understanding

Specification

- *Theories of moral understanding (Kohlberg) and/or prosocial reasoning (Eisenberg)*

The inclusion of the wording 'and/or' is important here, meaning that questions also need to contain that option. Therefore it is feasible to study either theories of moral understanding or theories of prosocial reasoning, or both.

Kohlberg's (1966) theory of moral understanding

Kohlberg's theory is based on cognitive development, seeing morality developing in a number of innate stages in a set order. It is seen as developing when biological maturation is sufficiently advanced, but disequilibrium plays a part, as experiences not fitting existing schemas challenge current ways of thinking about morality. Women are seen as less morally developed than men, as they are mainly restricted to a domestic life.

Each stage of morality is separate, because they involve a different kind of thinking to reach moral decisions, the focus being on how moral thinking occurs, rather than on what is thought about a particular moral issue. The theory believes that moral behaviour is a direct result of moral thinking.

Kohlberg created his theory as a result of ten moral dilemmas with no 'right' or 'wrong' answers, which he gave to participants. His initial study used 72 boys aged between 10 and 16 years of age, each one interviewed for 2 hours, with Kohlberg then assessing what stage of moral reasoning an individual was in.

One moral dilemma involved a chemist who will not give Heinz a drug that will save his dying wife, because he cannot immediately afford it. The moral dilemma is whether Heinz is right to steal the drug. Kohlberg is not interested in whether participants think it is right to steal, but in the reasoning behind their answer. From this he created three levels of morality, each one containing two stages (Table 7).

Table 7 Kohlberg's levels of morality

Level of morality	Description
1 Pre-conventional (age 6–13 years)	Stage 1: morality based on outcomes (e.g. punishments), rather than intentions
	Stage 2: moral rules followed when it benefits us
2 Conventional (13–16 years)	Stage 3: morality based on 'being good' and maintaining trust and loyalty of others
	Stage 4: morality based on what is best for society, fulfilling our duty
3 Post-conventional (16–20 years)	Stage 5: social order seen as paramount, with realisation that bad rules can be changed
	Stage 6: adherence to personal set of moral rules

Research

- Colby et al. (1983) tested the original sample for 26 years, finding at age 10 the majority showed stage 2 moral reasoning, with a few instances of stages 1 and 3. By age 22 the majority were in stages 3 and 4, with no one in stage 1. By the age of 36 the majority, 65%, were in stage 4, only 5% progressing to stage 5.
- Kohlberg (1969) tested the moral reasoning of participants in several cultures, such as Mexico and Taiwan, finding the same sequence of moral development. This suggests that transition through the stages occurs as an innate biological process.

- Kohlberg (1975) tested whether moral reasoning reflects moral behaviour. He gave students a chance to cheat on a test and observed that only 15% of participants with post-conventional morality cheated, while 70% of those with pre-conventional morality did, supporting the prediction.
- Rest (1983) performed a longitudinal study of 20 years' duration, following a sample of men from early adolescence to their mid-thirties. The stages of development were found to follow the set order proposed by Kohlberg, but change was found to occur very gradually. Most participants changed less than two stages, suggesting a degree of support for Kohlberg.
- Colby et al. (1983) reported that the moral dilemmas and interviews used to assess people's levels of morality made it impossible to differentiate between stages 5 and 6. Therefore stage 6 may be part of the normal developmental sequence after all.
- Fodor (1972) compared delinquents' and non-delinquents' levels of morality, finding non-delinquents at a higher level of morality. This supports the notion that moral thinking reflects actual moral behaviour.
- Walker et al. (1987) found that most children demonstrate stage 2 moral reasoning at age 10 and stage 3 by age 16, showing a progression supporting Kohlberg. However, nine rather than six stages were proposed, to cater for the fact that children often seem to be between stages.
- Berkowitz and Gibbs (1983) found that development through the stages was assisted by *transactive interactions,* where discussions are held about moral possibilities. This supports Kohlberg's idea that creating disequilibrium in an individual's way of thinking develops moral growth, but only if biological maturation allows.

Evaluation

- As he found no evidence of stage 6 reasoning in normal participants and little evidence of stage 5, Kohlberg (1978) decided that stage 6 might not exist.
- As only 12% of adults reach post-conventional morality, Atkinson et al. (1990) argued that it is more of a philosophical ideal, rather than part of a normal developmental sequence.
- Although Kohlberg's theory is not beyond criticism, it is supported by lots of research evidence.
- Hartshorne and May (1928) gave students opportunities to lie, cheat and steal and to spend money on themselves or others. They found that moral behaviour was situation-specific and not universal across all situations, casting doubt on Kohlberg's belief that moral behaviour is a reflection of moral thought.
- Moral dilemmas are not real-life scenarios and people may behave very differently from their moral reasoning if actually placed in such situations. Gilligan (1982) questioned women deciding whether to have abortions and found a different pattern of moral thought than Kohlberg, although this may be due to using female rather than male participants.
- Kohlberg's theory has been accused of gender bias. He saw morality as based on principles of justice, while Gilligan argues that women operate differently, on

principles of care. Kohlberg's negative rating of female morality is probably a result of participants being assessed by male-created standards, and of methodological faults, such as using only male participants.

- Although evidence suggests that the stages of moral development are cross-cultural, this may be because people of non-western cultures could not really appreciate dilemmas drawn from western cultural experiences, where individual needs are greater than those of others. Snarey et al. (1985) believes that the morality of collectivist cultures is centred on obeying elders and aiding society, yet these would be assessed at low levels by Kohlberg. This suggests that the theory may not be cross-cultural as it is making cultural judgements about morality.

Eisenberg's theory of prosocial reasoning

Eisenberg's theory has a broader approach, including the element of emotion not present in Kohlberg's theory, and centres on the idea of giving help and comfort to others. Similarly to Kohlberg, Eisenberg sees moral development occurring in tandem with development of general cognitive abilities.

A key feature of the prosocial approach is the idea of empathy — seeing from another's viewpoint. In this way, the emotions of others can be appreciated. For this reason, Eisenberg emphasised the use of role-taking skills, involving consideration of another person's perspective.

Similarly again to Kohlberg, Eisenberg used moral dilemmas in her research. These involved conflict between a person's own needs and those of others, with the influence of laws, rules, obligations, possible punishments, etc. at a minimum. A person has the option to bring help and comfort to others, but at a personal cost. Younger children make such decisions in a self-centred manner, while older children tend to take others' feelings into consideration. Children also show a progression of development as they age. Again like Kohlberg, Eisenberg used findings from her research to propose a theory of five levels (Table 8).

Table 8 Eisenberg's levels of prosocial reasoning

Level of prosocial reasoning	Description
1 Hedonistic (pre-school/early primary school)	Prosocial behaviour occurs when it benefits self.
2 Needs oriented (a few pre-school/ mainly primary school)	Some consideration of others. Little evidence of sympathy or guilt for self-centredness.
3 Approval oriented (primary school/ a few early secondary school)	Prosocial behaviour evident when it elicits praise from others.
4 Empathetic (late primary school/secondary school)	Some evidence of sympathy and guilt. Some reference to moral principles and obligations.
5 Strongly internalised (few primary school/some secondary school)	Evidence of internalised principles important to self-respect.

Research

- Eisenberg et al. (1983) performed a longitudinal study, following a group of 22 participants from age 4 to early adolescence, asking them questions about moral dilemmas. Consistent evidence was found to support her stage theory and the approximate ages therein. It was found that children who exhibit prosocial behaviour spontaneously at age 5 continue to do so in adolescence and early adulthood, suggesting that individuals possess consistent individual differences in levels of prosocial behaviour originating in childhood.
- During a follow-up of her sample, Eisenberg et al. (1991) found that the development of empathy plays a critical role in nurturing prosocial thought. Adolescents are likely to help others if they can empathise with their feelings.
- Caplan and Hay (1989) found that 3- to 5-year-old children demonstrated distress at another child's distress, but seldom attempt assistance, supporting the idea that empathy needs to be experienced for prosocial behaviour to occur.
- Hughes et al. (1981) found that children under 7 years of age could be distressed at others' sadness, but tended to explain this in terms of how it affected themselves. Children over 7 years of age could explain their distress in terms of the effects on others, suggesting that empathy cannot occur until sufficient maturation has occurred.
- Chalmers and Townsend (1990) found that girls with poor social skills were able to develop empathy and concern for others, if they received coaching in role-taking skills. This suggests that empathy can be learned and the use of role taking can facilitate this process.
- Midlarsky and Hannah (1985) found that younger children will help others, but for egocentric motives, such as wanting a reward or to avoid punishment. However, Roker et al. (1998) found that older children would help, but were more truly altruistic in their motives, because they understood the underlying principles of morality, supporting the idea of different levels of prosocial reasoning.
- Mills et al. (2004) examined potential sex differences in prosocial behaviour involving an element of self-sacrifice in participants aged between 17 and 68. Both males and females generally made the self-sacrificing choice, but there were gender differences in reasoning, with females showing more empathetic reasoning. This suggests that gender differences exist on a cognitive level, more than a behavioural level.
- Feshbach (1982) found clear evidence of females being more empathetic than males, while Aries and Johnson (1983) found females more likely to offer emotional support. However, Eisenberg (1986) reports that boys develop more slowly and become empathetic during adolescence.
- Eisenberg (1986) found evidence, from studies of European samples, of her levels of prosocial reasoning being cross-cultural, suggesting a link to biological maturation. However, evidence from collectivist cultures, such as from Kibbutz communes in Israel, indicates a lack of needs orientation, with reasoning more directed by communally based beliefs, suggesting that her theory does not fit non-western cultures.

Evaluation

- Theoretical support for the theory comes from Batson's (1991) empathy–altruism hypothesis, seeing human altruism being directed by the ability to sense the feelings of others. There is theoretical opposition, though: Cialdini et al. (1982) proposed the negative state relief hypothesis, perceiving assisting others who are distressed as being motivated simply by the desire to alleviate our own distress.
- The theory has practical applications, such as the training of parents, and others involved in childcare, in how to encourage and develop empathy in children: for example, getting them to look after pets or share toys.
- Although Eisenberg's theory is broader than Kohlberg's, incorporating the important element of emotion, there are many similarities and common themes, so it can be considered a development of Kohlberg's theory, rather than a separate theory.
- Eagly and Crowley (1986) believe that gender differences in empathetic concern can be explained by females traditionally developing nurturing and caring skills. For example, traditional female roles such as mother or nurse require nurturing and caring behaviour.
- The methodology used in Eisenberg's research is more valid than Kohlberg's, as her moral dilemmas were more suited to use with young children.
- It is possible that the development of empathy can occur earlier than Eisenberg stated. Zahn-Waxler et al. (1979) found concern for distressed others in children as young as 18 months.
- It is generally agreed that Theory of Mind (see the next section) appears in childhood, although there is disagreement as to whether this occurs suddenly at age 4, or occurs more gradually.
- There are ethical considerations, especially of harm, when researching on young children and exposing them to potential distress.

Development of social cognition

Specification

- *Development of the child's sense of self, including Theory of Mind (Baron-Cohen)*
- *Development of children's understanding of others, including perspective taking (Selman)*
- *Biological explanations of social cognition, including the role of the mirror neuron system*

The specification centres on explaining how individuals develop the ability to make sense of their social world. Theory of Mind, perspective taking and the mirror neuron system are all explicitly named and so must form part of your studies, as it is quite feasible for a specific requirement to address them directly in examination questions, either singularly or in unison with each other.

Development of the child's sense of self

Self-recognition

Having a sense of 'self' is an important mediating factor in social interactions. The ability to self-recognise is found in few animal and bird species, and is associated with the types of intelligence required for social living.

The standard test for self-recognition is the *mirror test*, where a coloured mark is made on the face and the participant placed in front of a mirror. If the participant then touches the mark, they are perceived as being able to self-recognise. Some infants can do this at 15 months of age, and the majority by 2 years. The rationale behind the test is that, for an individual to comprehend whom the mirror image is of, there must be a mental representation of self.

Research

- Amsterdam (1972) tested 88 children aged between 3 and 24 months, finding 42% of participants aged between 18 and 20 months and 63% of those aged between 21 and 24 months could self-recognise.
- Bertenthal and Fischer (1978) examined self-recognition from a Piagetian respective, noting the sensorimotor operations performed by children of different ages when regarding their own reflections. From this they devised five stages of self-recognition through which an infant should progress in its first 2 years of life, culminating in being able to name the image. The results supported the predicted developmental sequence of mirror-image responses. However, some of the criteria used have been criticised as not relating to self-recognition, such as touching a hat on your head when regarding your image.
- Lewis and Brooks-Gunn (1979) observed children between 9 and 12 months of age smiling at their own image, but not touching the mark on their face. By 21 months of age, 70% were touching the dot. The fact that children of 18 months identified with their own image in a photograph suggests that self-recognition is also possible from still images.
- Mans et al. (1978) showed that self-recognition in Down's syndrome children was delayed, but that by 4 years of age 89% could self-recognise, suggesting that recognition of self and thus self-awareness are related to cognitive development.

Gender concept

Development of gender identity is an important part of self-concept, most children developing **gender identity**, where they realise that they are male or female, between 2 and 3 years of age. Between 3 and 7 years, **gender stability** is developed, where it is realised that gender is fixed, and by 12 years of age **gender consistency** is achieved, where it is realised that changes in appearance or activity do not alter one's gender (for details, see 'Gender' in *Topics in Psychology* (1) for details).

The ability to differentiate between the physical self and the psychological self is seen as an important part of self-awareness, demonstrating awareness of a private, unseen self. Children between 3 and 4 years of age are aware of the distinction, but mainly refer to themselves in physical terms.

Research

- Eder (1990) found that 3- to 4-year-olds can describe how they behave in various situations, which is seen as an indication of their having awareness of a psychological self.
- Selman (1980) believed that children under the age of 6 could not distinguish between physical and psychological selves. By the age of 8 most children had developed the ability.

Self-referential emotions

Certain emotions, like embarrassment, convey a sense of self-awareness, involving thinking about yourself in relation to others.

Research

- Lewis et al. (1989) found that when children are asked to dance in front of adults, they display embarrassment around the same age that they demonstrate self-recognition.

Self-esteem

The ability to self-evaluate is dependent upon assessing the difference between the actual and the ideal self. It differs between individuals and across situations.

Research

- Harter (1987) found that a child's self-esteem is dependent on levels of perceived competence, reducing in mid-childhood due to input from others.
- Vershueren et al. (2001) found that children between 4 and 5 years of age have a sense of self-esteem and it is related to attachment patterns, with securely attached individuals having higher levels.

Theory of Mind (ToM)

Comprehension of another's thoughts and emotions is found in few animal species and is perceived as indicating higher intelligence. It can be tested for with the **Sally-Anne test**, assessing whether there is a realisation that another person has a false belief. Generally 4-year-olds can do this, although autistic children find it difficult, which may explain their difficulty in conducting social relationships.

The popular view of ToM is akin to Piaget's idea of egocentrism, with ToM seen as not developing until cognitive development does at around 4 years of age. A more recent view emphasises modularity, where specific brain areas, such as the amygdala and basal ganglia, are associated with ToM processing, with a set sequence of development and with ToM reasoning being inferred from other knowledge.

With the development of ToM comes the ability to manipulate and deceive by hiding one's emotions and intentions. This occurs from 3 years of age. It is possible there is a more primitive precursor to ToM, called Shared Attention Mechanism (SAM), developing between 9 and 18 months of age, allowing two people to realise they are attending to the same thing.

Research

- Harris (1989) reports that once children become aware of their own emotions, they can use them to pretend to be someone else, allowing the ability to be aware of others' thinking to develop at around 4 years of age.
- Shatz et al. (1983) believes children under 4 years of age can differentiate between different mental states. At 2 years of age they can name emotional states and by 3 years of age demonstrate knowledge of what thinking is.
- Flavell et al. (1986) found 3-year-olds who handled a sponge looking like a rock, called it a rock, while 4-year-olds called it a sponge, suggesting the development of ToM requires an appreciation of what is false, supporting the cognitive deficit explanation.
- Bartsch and Wellman (1995) found ToM acquisition follows a common developmental pattern in both American and Chinese children, suggesting a link with biological maturation.
- Frith and Frith (1999) found the amygdala, basal ganglia, the temporal cortex and frontal cortex show heightened rates of activity when participants had to consider others' mental states, supporting the modular view of ToM.
- Avis and Harris (1991) found that children in developed and non-developed countries realise at 4 years of age that people can have false beliefs, supporting the idea of biological maturation.

Evaluation

- Some studies of self-recognition, such as Dickie and Strader (1974), who put red tape on children's faces, may have provided tactile clues and may be regarded as lacking in validity.
- Some studies of self-recognition have accepted image naming as sufficient evidence for recognition of one's own image. However, this may be a learned response and not demonstrate the ability to self-recognise.
- A methodological difficulty in researching the development of social cognition is the lack of language skills in small children, making conclusions difficult to arrive at with any degree of certainty.
- Both visual self-recognition and self-referential emotions demonstrate the development of self-awareness.
- A child's sense of self may be affected by different factors. Case (1991) reported that 'I' comes from learning how behaviour affects others, while 'me' occurs as a result of watching your own movements.
- As children age, there is a gradual shift from describing the self in terms of activities, to describing the self using psychological characteristics. Hart et al. (1993) found this change most noticeable between 6 and 8 years of age.
- Damon and Hart (1988) found that the self-concept becomes increasingly socially based with age.
- Input from others seems to be important in determining self-esteem.
- It is generally agreed that ToM appears in childhood, though there is disagreement whether this occurs suddenly at 4 years of age, or more gradually.

Development of children's understanding of others

With the development of a sense of self comes the ability to understand others. The ability to self-recognise occurs simultaneously to acquiring empathy, at around 18 months of age.

Perspective taking

This concerns the ability to assume another's perspective and understand their thoughts and feelings. The understanding of others and yourself is enhanced by being able to differentiate between other people's perspectives and your own. Selman (1980) proposed **role-taking theory** to explain this development, where adopting the perspective of another allows comprehension of their feelings, thoughts and intentions. The theory was developed through research involving **interpersonal dilemmas**, such as whether expert tree-climber Holly, who had promised her father she would not climb trees any more, should climb a tree to save a kitten. The theory has five levels, developed from children's answers to question about the dilemmas (Table 9). As children mature, they take more information into account, coming to realise that people can react differently to the same situation. They develop the ability to analyse various people's perspectives from the viewpoint of an objective, neutral bystander and have a realisation of how different cultural and societal values affect the perception of the bystander.

Table 9 Selman's stages of perspective taking

Stage of perspective taking	Description
Undifferentiated (age 3–6 years)	Recognition that self and others have different thoughts and feelings, but often confuse the two
Social informational (5–9 years)	Recognition that different perspectives arise, as people have access to different information
Self-reflective (7–12 years)	Can perceive others' feelings, thoughts and behaviour from their perspective
Third party (10–15 years)	Can step outside a two-person situation, viewing it from a third-party neutral viewpoint
Societal (14–adult)	Understands that third-party perspectives can be influenced by societal value systems

Research

- Kravetz et al. (1999) used Selman's methodology to compare 22 normal children and 22 children with learning disabilities, on levels of interpersonal understanding, finding that severity of learning disability was positively correlated with difficulties in interpersonal relationships.
- Underwood and Moore (1982) found that the ability to perspective-take positively correlated with prosocial behaviour, suggesting that perspective taking enhances social relationships.
- Schultz and Selman (1990) found that the transition from self-centred perspectives to an ability to perceive from others' perspectives is related to the

development of enhanced interpersonal negotiation skills and concern for others, suggesting that perspective taking plays a key role in social maturation.

Evaluation

- The theory has practical applications as a means of conflict resolution. Walker and Selman (1998) used perspective taking to reduce violence levels by getting individuals to empathise with other people's feelings and viewpoints.
- Wentzel (1993) reports that perspective taking increases prosocial behaviour, in turn enhancing academic motivation and success.
- Schonert-Reichl et al. (2003) found that encouraging perspective taking in children reduced incidents of bullying.
- A lot of research into perspective taking is correlational and so does not show causality. Other mediating factors may be involved.
- Selman's stage theory and use of dilemmas provides researchers with an objective means of assessing social competence.
- Selman's research is used to ascertain when children are able take part in competitive sports, on the basis that competition becomes meaningful and positive when a child can conceptualise competition from others' viewpoints.

Biological explanations of social cognition

The search for a biological basis to social cognition is a recent one. It is thought that biological factors, such as genes and brain processes, interact with environmental variables, producing individual differences in social competence. Humans are extremely social animals and social interactive success is heavily dependent upon the development of brain systems geared to processing social information. Although learning experiences are necessary for normal development, without innate neural systems processing social stimuli, it is difficult to explain the universality and speed of social learning. Brain abnormalities impair social interactions in different ways, suggesting that social skills are directed by brain systems. Brain imaging studies indicate that a network of brain areas, linking the medial prefrontal and temporal cortex, form the neural substrate of mentalising, allowing representation of your own and others' mental states.

A **mirror neuron** is a nerve in the brain, active when a specific action is performed or is observed in another, allowing the observer to experience the action as if it were theirs. This allows us to share in the feelings and thoughts of others by empathising with and imitating others, and to have a ToM.

Research

- Rizzolato et al. (1996) found evidence of mirror neurons in the frontal and parietal lobes of macaque monkeys. These neurons behaved in the same manner when observing another monkey pick up food, as when the monkey did it itself.
- Rizzolato and Craighero (2004) used brain scanning on human participants, finding a network of neurons in the frontal and parietal brain areas appearing to work as mirror neurons.

- Gallese (2001) used fMRI scanning to find that the anterior cingulate cortex and inferior frontal cortex are active when an individual experiences emotion or observes another experience the same emotion, suggesting mirror neuron type activity.
- Wicker et al. (2003) found that smelling an odious aroma activated the anterior cingulate and insula brain areas, but observing another's facial expression of disgust also activated these brain areas, suggesting that mirror neuron activity allows empathy with others' emotions.
- Keysers et al. (2004) found that within the cortical region there is a localised neural network, activated when an individual is touched or when another person is observed to be touched, indicating possible mirror neuron activity.
- Stuss et al. (2001) reported that individuals with damage to their frontal lobes often had in an inability to empathise with and read other people's intentions and were easy to deceive, suggesting a biological link to social cognition.

Evaluation

- There is a methodological problem in studying mirror neurons in humans: it is not possible to study the actions of single neurons.
- Social cognition seems to exist only in some higher animals, suggesting it has a biological basis, which has evolved due to its adaptive advantage.
- Research indicates that it is the sensory-motor system that facilitates imitation of others, allowing empathy of observed emotions through simulation of the related body state.
- Dysfunction within the mirror neuron system may explain the inability of autistics to empathise with others.
- Jacob and Jeannerod (2004) have pointed out that a mirror neuron system is too simplistic an explanation — it cannot explain how the same actions performed in others can be interpreted differently by an observer in different contexts.

Questions
&
Answers

This section contains sample questions in the style of Unit 3. Each question is accompanied by guidance explaining the question's requirements, followed by a sample answer and examiner's comments/marks, detailing the strengths and weaknesses of each answer and explaining how the marks were awarded. Examiner's comments are preceded by the icon 𝒆.

Biological rhythms

(a) Outline research into biological rhythms. (9 marks)

(b) Evaluate the consequences of disrupting biological rhythms. (16 marks)

📝 Part (a) of this question requires descriptive material. Providing evaluation would not earn credit and would also waste examination time better spent elsewhere. The term 'research' relates both to research studies and to theoretical aspects. Most answers are likely to focus on the role of endogenous pacemakers and exogenous zeitgebers, as they are specifically named in the specification.

In part (b) you must avoid providing descriptive material, such as research studies into shift-work and jet lag, as this would not gain credit. What is needed is an assessment of the consequences of disrupting biological rhythms, such as explaining how the seemingly harmful effects of jet lag and shift-work may be due to other factors, or detailing practical applications. Commentary on ethical concerns would also be a relevant issue for inclusion.

■ ■ ■

Candidate's answer

(a) Siffre spent 6 months in a cave in Texas without natural daylight, developing a circadian sleep–wake cycle of just over 25 hours. He was connected to measuring equipment so his bodily functions could be monitored. He was able to eat and drink when he chose. He emerged after 179 days, thinking 151 had passed. When awake he had artificial lights on and this may have affected the validity of the study. This was a case study of one participant and may not be representative.

Aschoff and Weber found that participants living in an underground bunker with no natural light source had a circadian sleep/wake cycle of about 26 hours. This was similar to what Siffre found, showing that endogenous pacemakers can control this circadian rhythm in the absence of light cues, but that light does seem necessary to synchronise the internal biological clock with the external environment.

Russell found that if underarm sweat of a woman was smeared on to the upper lips of other women, their menstrual cycles became synchronised. The sweat was collected on a pad and treated with alcohol to destroy bacteria.

McClintock and Stern found that armpit swabs taken from women about to ovulate made women who sniffed them have shorter menstrual cycles, but swabs taken from women who had recently ovulated made other women have longer menstrual cycles, suggesting the pheromones in the sweat affected their infradian rhythms.

(b) Research supports the idea that consequences of jet lag are more severe when travelling on eastbound flights than westbound. Klein found adjustment easier for participants on westbound flights and this was not to do with whether it was the outbound or homebound flight, therefore suggesting phase advance has more severe consequences than phase delay.

Schwartz backed this up by finding that baseball teams from the west coast of the USA had a worse win–loss record playing teams from the east coast than the west, suggesting that phase advance has more severe consequences. However, another explanation for the results is that teams from eastern states are superior baseball teams.

Webb and Agnew found that the negative effects of jet lag could be addressed by sticking to regular meal times of your new time zone and by exposure to natural light. These appeared to help re-set body clocks, showing how psychological research can lead to successful practical applications, with another possibility being melatonin supplements.

It is suggested the desynchronisation effects of working irregular shifts were responsible for the nuclear accidents at Three Mile Island and the Chernobyl nuclear plant, implying that the concentration lapses associated with shift-work may be responsible for decision failures that occurred. Also the fact that shift-work causes disruption to concentration suggests that shift-work has a cognitive effect.

Research by Colligan suggests that the consequences of shift-work rotations are generally negative. Workers had more minor illnesses, digestive problems, drank more and had less successful relationships. However, Czeisler came up with a solution to this. He persuaded bosses to delay shift-rotations so workers had time to biologically adjust, and to move to a phase delay system of always rotating shifts forward in time. Workers became healthier and happier and output increased, demonstrating how research can lead to successful practical applications.

Overall, research into disrupting biological rhythms shows how nature, in the form of innate endogenous pacemakers, interacts with nurture, in the form of environmental zeitgebers.

This is a good answer. In part (a) four relevant studies are offered, two concerning circadian rhythms and two the infradian menstrual cycle, and these are generally accurate and reasonably detailed, putting the answer in the 'reasonable' mark band. The question only calls for descriptive material, but some evaluation is included, such as case studies not being representative and the suggestion that pheromones affect infradian rhythms. These comments do not earn credit, although no marks are 'taken away' — with positive marking, credit can only be gained and not lost. Similarly, the non-inclusion of dates of research incurs no penalty.

In part (b) it would have been quite easy to fall into the trap of describing the consequences of disrupting biological rhythms. However, such material would be non-creditworthy as marks are only available for an evaluation of the consequences.

Fortunately, that is what this candidate gives, ensuring that the material used is relevant by using wording such as 'research supports' and 'this suggests'. Good use is made of possible outcomes of research, such as practical applications. There is inclusion of material relating to the nature–nurture debate and a consideration of other interpretations of research data. Overall, the answer is best suited to the requirements of the 'reasonable' mark band, and at the top end because the answer is closer to being effective than basic. For a better mark, more coherent elaboration is needed and greater structure is required to the ideas expressed.

(AO1 = 6/9) + (AO2/AO3 = 12/16) = 18/25 marks

uestion

Sleep states

(a) Outline two theories of the functions of sleep.   (9 marks)

(b) Evaluate one of the theories you have outlined in (a).   (16 marks)

For part (a) the evolutionary explanation and the restoration theory are both specifically named on the specification and so would form the basis of most answers, although any other theory dealing with the functions of sleep is equally acceptable.

Part (b) clearly states that only one of the two theories offered in (a) has to be evaluated, so care must be taken to not evaluate both. Alternative theories could be introduced and earn credit if used as part of a sustained and effective commentary, although the focus must be centred on the selected theory. Supporting evidence from research studies could provide effective evaluation, as would the explanatory value of the theory.

■ ■ ■

Candidate's answer

(a) The evolutionary explanation sees sleep serving an adaptive advantage, with different species evolving different patterns of sleep to deal with their specific environmental requirements.

Meddis (1979) believes that sleep evolved to keep animals safe from predators when their necessary tasks, such as foraging, are not required, meaning that prey animals tend to sleep less, as they have to be vigilant.

Webb (1982) sees hibernation as an evolutionary response to harsh times of food scarcity and extreme weather. Hibernating animals can conserve energy and stay at a regular temperature.

Aquatic mammals have evolved strategies allowing them to keep breathing, like only one brain hemisphere sleeping at a time, or taking lots of micro sleeps.

Smaller animals usually sleep more to prevent their active metabolism reducing energy stores quickly.

The restoration theory sees sleep as being a rest period allowing repair and rejuvenation of body and mind.

Horne (1988) thinks stage 4 and REM sleep allow the brain to refresh itself, so cognitive functioning is at required levels.

Oswald (1980) states that growth hormone is produced, along with other hormones, during the four stages of NREM sleep, allowing the body to physically repair and restore itself. Growth hormone stimulates tissue growth, assisting the synthesis of proteins necessary for repairing damaged tissue.

(b) One way of testing evolutionary theory is to look for examples in the natural world that fit predictions. For example, research backs up the idea of aquatic animals evolving sleep strategies protecting them from drowning. Pilleri (1979) found that Indus dolphins have frequent micro sleeps, while Mukhametov (1984) found that the bottlenose dolphin has one cerebral hemisphere asleep at a time. However, there are also real-life examples not fitting evolutionary theory — for instance, giant sloths sleep for 20 hours a day and yet have a large body size and a relatively slow metabolism. Another cautionary point is that a lot of research into evolutionary explanation involves animals, causing problems in generalising to humans, though the fact that sleep is universal does suggest it has evolved to have some adaptive purpose. To be regarded as ethical, studies on wild animals should not reduce their fitness in any way.

Stear supports the evolutionary explanation because he believes sleep conserves energy when an individual does not need to be lively and therefore different species evolved different sleep patterns as an adaptation to their specific ecological needs. This is backed up by real-life examples — for instance, prey animals like gazelles do not sleep a lot, as they need to spend lots of time grazing, while lions sleep a lot because they only need to eat about once every 5 days. However, prey animals are usually herbivores. Human sleep patterns may suit ecological needs in the EEA but do not fit the modern world. Perhaps the fact that humans now sleep less means our sleep patterns are slowly adapting.

The evolutionary explanation is regarded by some as reductionist, because it reduces the complex behaviour of sleep down to the single factor of adaptiveness. It can equally be seen as being deterministic, perceiving sleep as behaviour coded into our biology, with no role for free will.

This is an excellent answer. The candidate focuses on two relevant theories and, through the use of a concise writing style, creates an answer with wide-ranging breadth and a decent level of detail. The evolutionary explanation is well covered in a coherent and clear manner. The explanation of restoration theory lacks a little at times: for example, it does not explain how REM and stage 4 sleep allow the brain to refresh itself.

The evaluation demonstrates sound analysis and understanding. There is a clear focus and coherent elaboration throughout, with ideas well structured and clearly expressed. Evaluative points are built upon each other to construct an effective commentary, with elements of ethical and methodological considerations also apparent. The line of argument could possibly be a little clearer towards the end.

(AO1 8/9) + (AO2/AO3 = 15/16) = 23/25 marks

Disorders of sleep

Discuss explanations for insomnia. (25 marks)

The term 'discuss' means that there is a requirement to describe and evaluate, in this particular case, explanations for insomnia. Material could be included on both primary and secondary insomnia, as well as factors influencing insomnia, such as personality. Support from research findings would provide a good source of evaluative material, as well as an assessment of the methodological problems in researching into this area and the inclusion of practical applications that help sufferers. In order to gain access to the higher mark bands, all Unit 3 questions require reference to debates and approaches. Therefore the inclusion of relevant material concerning the biological approach could perform that function here.

■ ■ ■

Candidate's answer

Insomnia is a sleep disorder where sufferers have problems initiating and maintaining sleep. One factor linked to insomnia is sleep apnoea, a medical condition where people have persistent pauses in their breathing, sometimes for minutes, culminating in a loud snort when they recommence breathing. This can happen up to 200 times a night. There are two types, obstructive sleep apnoea, caused by blocked airways, common in over-weight people, and central sleep apnoea, occurring due to faulty signalling to brain areas linked to breathing.

Sleep apnoea is more common in older adults and will become more of a problem, because the population is ageing, creating a need to find treatments for the condition. This notion is backed up by Morrell (2000) who found that up to 20% of older adults have the condition, a rate ten times that of younger adults, though the condition tends to be more severe in the young. Clinicians report the condition on the increase among younger sections of the population and this is linked to the growth in obesity. It is thought that the difference in prevalence rates may be due to physical changes in the cardiovascular system as we age. One implication of this is that different forms of treatment may be required for people of different ages, although addressing the problem of obesity would seem to be a universal solution.

Stickgold (2009) sees insomnia as causing mental problems like depression; apnoeic insomniacs have twice the rate of the normal population. However, Horne (2009) thinks it more likely that mental disorders lead to insomnia.

Another factor associated with insomnia is personality. Especially implicated is psychasthenia, a personality disorder similar to obsessive–compulsive disorder, where a person has fears, doubts, anxieties and obsessive compulsions. Other personality factors have been identified, like heightened emotional arousal.

Lundh (1995) tested 233 persistent insomniacs, finding a high incidence of psychasthenia, with sufferers also low on self-esteem and over-dependent on others. This supports the idea that personality plays a role in causing and maintaining insomnia.

Kales (1976) personality tested 124 insomniacs, finding 85% had abnormal personalities, with a high incidence of psychasthenia, depression and conversion hysteria. Sufferers were highly emotionally aroused because of a tendency to internalise psychological disturbances, giving backing to Lundh's findings and implying that a psychophysiological mechanism may be responsible for insomnia.

A good means of providing proof that personality traits can lead to insomnia would be to conduct longitudinal studies on people identified as being at risk from their personality profiles. There is a danger, though, of creating a self-fulfilling prophecy, which in turn could create an ethical problem of causing psychological distress. Research suggests the best way to combat insomnia in such cases is to address the personality abnormalities rather than the insomnia directly, indicating that it is personality causing insomnia.

Grano (2006) found male insomniacs are often impulsive, suggesting a gender difference in associated personality traits.

Explanations of insomnia show how biological factors interact with psychological factors to cause problematic sleep disorders.

This is an excellent answer. The candidate demonstrates a sound knowledge and understanding of the subject area, and a good range of relevant material has been selected. Two explanations are described accurately with breadth and depth of detail.

There is substantive evidence of coherent elaboration, with several threads of thought being used to construct an effective commentary. Ideas are well structured, with relevant methodological points being well utilised. The candidate appears to run out of time at the end, finishing with a couple of unconnected points.

(AO1 = 9/9) + (AO2/AO3 = 14/16) = 23/25 marks

Question 4

Formation, maintenance and breakdown of romantic relationships

(a) Outline two explanations relating to the breakdown of relationships. (9 marks)

(b) Evaluate one of the explanations relating to the breakdown of relationships that you have outlined in (a). (16 marks)

✍ For part (a), there are several ways of explaining the breakdown of relationships. Although not specifically named in the specification, the stage models of Duck and Lee would prove a popular inclusion in many candidates' answers. However, more theoretical accounts, such as the equity model, are also acceptable, as long as the focus of the answer remains upon relationship breakdown. If more than two explanations were given, all would be marked, but only the best two would be credited; and if only one explanation were provided, this would be considered partial performance, restricting the number of marks available.

For part (b), evaluation could be achieved in terms of support from research evidence, or maybe through comparison with alternative explanations. Material on gender and/or cultural bias would also be relevant as long as explicitly linked to the explanation presented.

■ ■ ■

Candidate's answer

(a) One explanation relating to the breakdown of relationships is Duck's. He showed breakdowns occur in four stages. Firstly there is the 'intra-psychic' stage where a person mentally thinks about the negative aspects of a relationship and tries to resolve them and if not, they move to the next stage, the 'dyadic stage', where the other partner in the relationship comes to know about what is thought in terms of how successful and fulfilling their relationship is. If they can sort out the problems they will, if not they move to the next stage, the 'social stage' where the break-up becomes public and there is negotiation about children, property and finances. This is a time when family and friends support the couple into rebuilding their relationship. If not the process moves to the final stage, the 'grave dressing' stage where the couple work on rebuilding their life after the break-up, giving their own account of the relationship breakdown. This helps an individual to accept what has happened and makes them more suitable for another partner.

Lee provided another explanation for the breakdown of relationships. He stated that people break up because of internal factors. When two people spend a lot of time together, a person notices qualities they may find distasteful, such as boredom and lack of communication. External factors also play a part, like opinions of friends and families and incompatible working hours.

There are also five stages relating to why the relationship breaks down. Firstly there is 'dissatisfaction' where a person in the relationship is unhappy, then 'exposure' where the other partner finds out how the other person feels. After this comes 'negotiation', where both partners try to work on their relationship and improve it. If they're unable to do this they move to the next stage, resolution, where both partners decide to end the relationship. After the relationship has ended it reaches termination, both partners going their separate ways.

(b) Lee's explanation doesn't take into account positive aspects like the couple resolving their issues with counselling. In comparison to Duck's model, Lee's is more successful. However, if they were combined together it would give a more realistic view of why the relationship has actually ended.

Lee's model doesn't assess the feelings of each partner in a relationship when the break-up is occurring — for example, one person may be dwelling on the negative aspects, while the other person thinks there are no faults at all.

The explanation doesn't suggest what happens after the break-up, as there must be a need to resolve other issues.

Individual differences aren't taken into account as everyone may have a different form of break-up, so these explanations are not universally applicable.

This is a reasonable answer. Part (a) is done well. Two relevant explanations, those of Duck and Lee, are outlined accurately, clearly and with appropriate detail. Duck's explanation is probably done best as Lee's explanation lacks a little depth at times. Unfortunately the candidate has a long-winded writing style and although the content is fine, the candidate takes a long time outlining it, reducing the amount of time that can be dedicated to part (b), where the majority of the marks are on offer.

Part (b) is too short, probably because the candidate has dedicated too much time and effort to part (a). This is a common occurrence and is often the determining factor between those candidates who get high marks, and thus good grades, and those who do not. This might happen because people feel they have to say everything they can think of in their description/outline. Remember, you only have to provide enough material to earn the 9 marks on offer. This is a skill you can practise by constructing sample answers as a regular part of your learning. It might also occur because some people do not understand what is meant by evaluation. Again this is something that can be practised until you get it right.

With this particular answer, although some pertinent points are made and the comparison with Duck's theory often works well, there is no use of research evidence, comparison with other cultures or indeed comparison with other explanations. Some of the points made are also unclear, such as why combining Duck's and Lee's theories gives a more realistic view of why the relationship ended.

(AO1 = 8/9) + (AO2/AO3 = 6/16) = 14/25 marks

uestion

Human reproductive behaviour

Critically consider the relation between sexual selection and human reproductive behaviour.

(25 marks)

AO1 credit could be gained by describing a range of human reproductive behaviours related to sexual selection, such as parental investment and mate choice. It is important to remember that the provision of material on non-human animals could only earn credit if explicitly linked to human reproductive behaviour.

As the majority of marks, 16 of them, are available for evaluation, the majority of time and effort should be concentrated on that part of the answer. The degree of support from research studies could help to achieve that aim, as well as the provision of relevant methodological criticisms. Cross-species comparisons could also prove effective, as well as a consideration of cultural differences in human reproductive behaviour.

■ ■ ■

Candidate's answer

Sexual selection involves selecting characteristics for reproductive success, the prime aim being to produce healthy children who go on to have reproductive success themselves. Towards this end, male and female reproductive behaviour is different, due to different selective pressures upon them. Males are never certain of the paternity of children and produce lots of sperm over a relatively long time and potentially can fertilise many females at little cost to their reproductive potential. A male's best strategy is to maximise the number of possible couplings and this leads to intrasexual competition between males and polygamy, a system where a male mates with several females, preferably who exhibit signs of fertility.

Females are certain of the maternity of children, but only produce a few eggs over a shorter period, each egg representing a sizeable chunk of her reproductive potential. She also has to bear the costs of pregnancy and subsequent childcare. A female's best strategy is to indulge in intersexual competition, by choosing partners who advertise genetic fitness. This can occur through lengthy courtship rituals that encourage males to invest time and resources into the potential relationship, increasing the chances he will not desert and will produce further resources and protection.

Buss (1993) found that men are fearful of a female being sexually unfaithful, while women fear men being emotionally unfaithful, illustrating the differences between male and female reproductive behaviour, as males fear being cuckolded and raising a child not genetically theirs, while females fear males spending resources on another female.

Two further pieces of research that show how sexual selection is related to reproductive behaviour come from Boone (1986), showing that females prefer older males who are resource rich, and Kenrick and Keefe (1992), who found that males prefer younger females. This again demonstrates the differences in male and female strategies.

The 'sexy sons' hypothesis demonstrates the relationship between sexual selection and reproductive behaviour. Females select attractive mates so their sons will have the same qualities, increasing reproductive potential. These qualities become favoured by natural selection, but become so exaggerated that they become bizarre, like the intricately decorated bowers that male bowerbirds produce to attract females.

Partridge (1980) gave some female fruit flies a free choice of partners, while others were forcibly mated with random males. The offspring from the free choice matings were superior in their competitive ability, suggesting that females can improve the reproductive potential of their children by selecting partners with good genetic quality. There has to be caution in generalising animal findings on to humans though.

However, evidence from human studies supports those from animal studies. For example, Cerda-Flores (1999) investigated the theoretical male strategy of sneak-copulations, where mating with non-partner females wherever possible increases reproductive fitness. It was found that 12% of children in Mexico were fathered outside of regular partnerships, lending support to the concept. However, Sasse (1994) found a figure of only 1.4%. This could be due to cultural differences, or differences in types of sample — for instance, using DNA data from examples where males had suspicions about the paternity of their children.

Evolutionary theory sees male and female sexual selective behaviour in terms of maximising reproductive potential, but these behaviours can be explained in other ways — for example, gender socialisation can account for female selectiveness and male promiscuity.

One final problem is the difficulty in identifying and separating out the effects of sexual selection from natural selection, making it difficult to perform valid research.

🖉 This is an excellent essay in both its description and its evaluation. The candidate adopts the sensible tactic of describing the relationship between sexual selection and human reproductive behaviour first, allowing the candidate to monitor the amount of material outlined, and how much time this has taken. The differences between male and female strategies and the reasons for them are described accurately, clearly and with a good range of breadth and detail. Another way of achieving this would be to outline the various male and female reproductive strategies, such as sperm competition and the handicap hypothesis. However, as this candidate has realised, you do not need to put everything in to earn all the marks available.

The evaluation uses a combination of research evidence and theoretical points, which are well elaborated and woven into an effective commentary. The candidate also shows an awareness of pertinent methodological issues as well as some of the limitations of evolutionary explanations.

(AO1 = 9/9) + (AO2/AO3 = 16/16) = 25/25 marks

Effects of early experience and culture on adult relationships

Discuss the influence of childhood experiences on adult relationships. (25 marks)

Providing descriptive material on the influence of childhood experiences on adult relationships — for example, the continuity hypothesis — earns AO1 credit. The specification also lists adolescent experiences, but as this question is specifically focused on childhood experiences, the inclusion of descriptive material on adolescent experiences would not be creditworthy.

Evaluation for AO2/AO3 credit could include the degree of support from research studies and relevant methodological concerns, as well as the problem of determining whether adult relationships are indeed affected by childhood experiences or by other factors and the difficulty in assessing overall developmental outcomes. The somewhat deterministic nature of attachment theories could be used to satisfy the need to include material on issues, debates and approaches.

■ ■ ■

Candidate's answer

Psychologists have been interested in seeing if the quality and types of adult romantic relationships are influenced by childhood attachments. Bowlby had a theory of attachment called monotropy where a child forms an emotional bond with its significant carer and he thought this attachment created the foundations for that individual's adult relationships. He saw this early maternal bond as an internal working model for the future and it is known as the continuity hypothesis, because it continues through life, suggesting it is important that we form warm, loving, secure bonds if we want to have successful adult relationships. Bowlby blames high divorce rates on poor-quality attachments in childhood, which has such a high cost for society. If true, it is important that psychologists develop strategies helping parents form secure attachments with children.

Margaret Ainsworth, using a methodology called the 'Strange Situation', which became the standard tool for measuring attachments, found there are several standard attachment types, secure, insecure-avoidant and insecure-resistant, and these are seen as giving a child a set of beliefs about themselves, others and the nature of relationships. Therefore someone fortunate enough to have a secure relationship should go on to have similar contented relationships as an adult.

Bowlby's theory was extended into the arena of adult relationships by Hazan and Shaver, who saw it as affecting romantic relationships, caregiving and sexuality. They devised a love quiz in a newspaper, getting readers to volunteer their feelings and experiences about romantic relationships and childhood relationships with parents.

A strong relationship was found between the two, supporting the continuity hypothesis. For example, those with secure childhood attachments were more likely to have lengthy adult romances, while those with insecure-resistant attachment types tended to be untrusting and had poor-quality adult relationships. Insecure-avoidant types didn't believe in love and feared closeness to others and so also had poor-quality, relatively short adult romances. As the initial participants were people who volunteered information to the newspaper quiz, it is unlikely that they gave informed consent, had the right to withdraw or were debriefed.

Simpson (2007) found evidence supporting the continuity hypothesis by performing a longitudinal study, finding that those with secure attachments were more socially competent and developed firm friendships. However, it is probable that other factors play a role, like the different attachment styles that people bring into a relationship.

Further evidence supporting the hypothesis comes from McCarthy (1999) who found that women with insecure-avoidant attachments in childhood had poor later adult relationships, while those with secure attachments were lucky in love, having successful romances and friendships.

However, an alternative explanation is the temperament hypothesis, which sees childhood attachments and adult relationships as being determined by innate, non-changing personality factors.

Meier et al. (2005) found that the type and quality of adult relationships related to that of adolescent relationships, suggesting a link.

Connelly and Goldberg (1999) supported these findings by finding that the degree of intimacy in young adult relationships was related to the degree of intimacy in peer relationships.

A general criticism is that attachment theories are deterministic, as they see childhood attachments as causing the type and quality of adult relationships. However, other factors play a part, like the interaction of attachment styles of the people in a relationship.

🖉 This is a good answer. The continuity hypothesis is well explained in terms of Bowlby's original theory and Hazan and Shaver's later adaptation of it into adult relationships. A fair level of generally accurate detail is provided, with some evidence of breadth and depth.

Research evidence is used in a reasonable fashion to assess the level of support for the hypothesis and there is also evidence suggesting an understanding of some of the relevant methodological issues.

Towards the end of the answer, the candidate wanders away from the question, offering material on the influence of peer relationships on adult relationships. This is not relevant and gains no credit; not that it loses any marks. The essay concludes on the relevant issue of determinism, although there is a degree of repetition of a point already made.

(AO1 = 7/9) + (AO2/AO3 = 9/16) = 16/25 marks

Social psychological approaches to explaining aggression

Outline and evaluate two social psychological theories of aggression. (25 marks)

✐ Although not specifically mentioned in the specification, social learning theory and deindividuation would probably form the basis of most candidates' answers, although any relevant theory, such as relative deprivation theory, would be equally acceptable. There is a specific requirement to outline two theories, so only providing one limits the number of marks available, while providing more than two results in only the best two gaining credit.

The theories could be evaluated in terms of support from research studies, although it is important to remember that merely providing a description of such studies, Bandura's Bobo doll study being a relevant example, does not earn AO2/AO3 credit. Moreover, simply detailing the methodological limitations of research is not generally considered a high-level form of evaluation. Comparison of the two theories in terms of strengths and limitations could form a more effective route to gaining evaluative credit, and the same end could be achieved by comparisons with non-social psychological approaches, or by making a general assessment of social psychological approaches.

■ ■ ■

Candidate's answer

One social psychological theory of aggression is social learning theory, which sees aggression as learned in two ways, both involving operant conditioning: first by direct means, where an aggressive behaviour is reinforced as it occurs, and second indirectly, where an observed aggressive behaviour is reinforced and then imitated. Therefore social learning teaches us how and when to imitate specific types of aggression.

Bandura outlined several steps in modelling aggressive behaviour. First, we have to pay attention and more attention is paid to attractive, high-status models, especially people we identify with. Second, observed behaviours need to be memorised if they are to be recalled and imitated. Third, imitation only occurs if we have the skills necessary to reproduce the behaviour — for instance, imitating someone with martial arts skills. Finally, the use of reinforcements creates the motivation for the aggressive behaviour to be imitated. Also a high level of self-efficacy, situation-specific confidence, is required for us to imitate an aggressive model.

Support for the social learning theory comes from Bandura's Bobo doll study, suggesting that aggressive acts will be observed and imitated if we see an aggressive model being reinforced for their actions, but will not if we see the model being punished. This suggests we learn aggressive actions through observation, but only replicate them in situations where we see them as being appropriate. Support for the theory's belief that we are more likely to imitate models we identify with, comes from the fact that children imitated more aggressive behaviour if the model was the same sex as them.

However, there are criticisms of the Bobo study, casting doubt on the theory's validity. First, there are methodological issues, like the fact that the doll is designed to be hit and the children were deliberately frustrated beforehand. Also it was the children who teachers rated as naturally aggressive who imitated more aggression, suggesting personality plays a part. Also is it ethical to encourage children to be aggressive? It could affect long-term behaviour.

The theory is able to explain why people's aggression levels vary between situations, because they have been reinforced that way. If aggression were biological, aggression levels would be consistent.

However, the theory does not consider biological factors, like levels of testosterone, and therefore is an incomplete explanation.

Another social psychological model of aggression is deindividuation, which occurs when someone loses their sense of identity. Normally people won't indulge in aggression, as they are easily identifiable, but in crowds there is less social restraint and personal responsibility is less, so we may behave aggressively. In such situations we may fail to follow our personal norms, instead following group norms and these may be aggressive ones.

Being in a crowd can make us feel anonymous, especially if our features are disguised. This leads to having a reduced awareness of self and a lesser sense of guilt, as well as a reduced fear of being caught and punished. Therefore the bigger the crowd, the more chance we will be deindividuated and behave aggressively.

Research evidence supports the idea of deindividuation. Malmuth and Check found a third of male students would rape if they could get away with it, while Zimbardo found, in a replication of Milgram's shock study, that teachers deindividuated by wearing a hood gave more shocks. Diener found that anonymous trick-or-treating children stole more sweets than non-anonymous children, suggesting that reduced self-awareness and heightened anonymity make people more likely to be anti-social, including being aggressive.

✍ This is a good answer. The description of both the social learning theory and deindividuation is accurate, clear and reasonably detailed, although the answer is slightly overlong and in need of a more concise writing style. One thing this candidate does well is to tie in the outlines to explanations of aggression, rather than just present a general outline of the theories.

One of the most important skills to develop when writing answers is that of balance: supplying enough material on the different aspects of a question to maximise the marks you can accumulate. This candidate unfortunately gets it a little wrong, because with the outlines being lengthy and the evaluation of the social learning theory also being overlong, little time is left to produce a full evaluation of the second theory. Only by practising writing answers like this will you learn to get it right.

The evaluation supplied is accurate, relevant and clearly expressed, and makes good use of research evidence. However, because the writing style lacks conciseness, there is a lack of depth at times.

(AO1 = 7/9) + (AO2/AO3 = 11/16) = 18/25 marks

Biological explanations of aggression

Compare and contrast social psychological and biological explanations of aggression. (25 marks)

This is an example of a question cutting across two subsections of a topic area. There are 9 marks available for AO1 and to earn credit here requires outlines of both the social and the biological explanations of aggression, drawing out the main points in doing so. Marks for evaluation would be gained, as the question suggests, by comparing and contrasting the two approaches in terms of strengths and limitations. The quality of research evidence on which both approaches are based could also be assessed: for instance, in terms of relevant methodological and ethical issues.

■ ■ ■

Candidate's answer

Social psychological theories of aggression see aggression as externally determined by social factors, while biological theories see aggression determined by internal physiological factors. For example, one social psychological explanation is social learning theory, which sees aggressive behaviour as being learnt, either by direct reinforcement, or vicariously, by imitating models seen to be reinforced. The more a reinforced model is identified with, the more they will be imitated. Bandura performed several studies where children, when given the opportunity, imitated an aggressive model seen to be reinforced by praise.

Biological models, on the other hand, see aggression as linked to genetics, hormones, neurotransmitters and brain structures. Low levels of the neurotransmitter serotonin are linked to heightened aggression and high levels to non-aggression. The male hormone testosterone is also related to aggressive behaviour — when given to females it can elevate aggression, and castration reduces testosterone and thus aggression levels. Testosterone has been linked to a critical early period in life, where exposure to testosterone is essential if it is to stimulate aggression in adulthood. It is also thought that testosterone helps to sensitise the androgen responsive system, so it is responsive to male hormones.

Several genes are linked to aggression, especially MAOA, which helps regulate levels of various neurotransmitters. Various brain structures have also attracted interest, especially the cortex's role in inhibiting the limbic system, with the amygdala attracting most interest.

A strength of social psychological explanations is that they can explain variations in people's levels of aggression in different situations as being due to varying levels of reinforcement. If aggression were purely biological, then aggression levels would be consistent across situations. However, social theories are impoverished, not taking into account biological factors, though Bandura did acknowledge that aggressive urges are biological, but argued that knowing how and when to be aggressive was socially determined. Therefore a better explanation is one perceiving biological factors as interacting with social ones to determine actual aggressive behaviour. Social learning of aggression can be seen as more powerful than biology, because it explains why different cultures have different levels of aggression, as being due to culturally determined learning experiences. For example, Mead found tribes in New Guinea, differing widely in aggression types and levels. In this same way social theories can explain individual differences in aggression levels, though biological theories can also explain them as varying levels of neurotransmitters, genes and hormones.

Social theories suggest practical applications, as reinforcements could be used to diminish aggressive behaviours and build up non-aggressive ones. If biological explanations are true, they too could provide practical applications in the form of gene and drug therapies. Those who see Raines' findings of abnormalities in the brains of murderers as suggesting a causal link to violent acts have even suggested pre-screening of people to identify those who should be locked up before they offend. However, the results are only correlational, suggesting that other factors are involved.

Any explanation seeing biology alone as determining aggression would be a deterministic one, seeing no role for free will. Although this may be true for lower animals, human behaviour generally involves cognitive and social inputs. Biological explanations can be seen as reducing explanations of aggression down to the single factor of biology and this is too simplistic, as research indicates that aggression is determined by an interaction of factors, both socially and biologically determined.

This is an excellent answer. Some of the content of this answer, especially relating to social theories, is similar to that of the previous answer. However, here a more concise writing style allows more detail to be included and a better balance between the different elements of the answer.

The candidate takes a mainly theoretical viewpoint, and as this is related to explanations of aggression, this is fine — what is being compared is two theoretical viewpoints. A concise and well-structured outline of the two approaches is evident and both are accurate and clear, with a good degree of breadth and depth, especially for the biological approach.

The evaluation is effective, drawing out positives and negatives of the two explanations, and there is a good element of the two being combined and compared to form an effective commentary. Relevant issues of determinism and reductionism are included. One way this answer could be improved would be by the inclusion of comments relating to methodological considerations.

(AO1 = 8/9) + (AO2/AO3 = 14/16) = 22/25 marks

Question 9

Aggression as an adaptive response

(a) Outline one explanation of group display in humans. (5 marks)

(b) Outline and evaluate the role of infidelity and jealousy as explanations of human aggression. (20 marks)

As part (a) is only worth 5 marks, care should be taken not to include too much descriptive material, as this approach would not earn any extra credit. Only one explanation is required, so candidates providing more than this requirement would have only their best explanation credited.

In part (b) a brief outline of the role of infidelity and jealousy is necessary, as the main bulk of the marks, 16 of them, are available for the evaluation. This could be achieved by considering whether research evidence backs up the idea that aggression is related to jealousy and infidelity. An assessment of whether gender and cultural factors are more important than innate factors in this area could also prove effective. The deterministic and reductionist nature of evolutionary theory could be addressed as a means of including material on issues, debates and approaches in order to gain access to the higher mark bands.

■ ■ ■ ■

Candidate's answer

(a) Group displays are ritualised forms of aggression often found at sporting events, serving to motivate, determine dominance hierarchies and intimidate other groups. War dances and supporter displays have been incorporated into sports events as group display vehicles, like the Siva Tua performed pre-match by Samoan rugby players, or football fans wearing face paint, which both motivate and intimidate.

Group displays at sporting events mark out territories to which other teams' supporters are denied access; indeed, many team sports involve the defence of home territory and the invasion of opposition territory. The ritualised element of sports events can be seen in the aggressive posturing and verbal abuse between sets of fans, rarely breaking into harmful violence.

(b) Although not the only explanation of human aggression, infidelity and jealousy play a significant part. Jealousy involves fear of losing affection or status and can lead to aggressive behaviours between males competing in ritualistic ways to be chosen by females. It is also seen in females, though they tend to get jealous and aggressive over other females' perceived levels of attractiveness. Men desire pretty girls, because beauty is an indicator of fertility and it is in a male's interest to mate with fertile females, because he will get his genes into the next generation. This is covered by the theory of evolution that sees behaviours having an adaptive advantage in meeting environmental demands, as being naturally selected and thus becoming more widespread throughout the population. Females tend to be 'choosy' because they have a lot of investment riding on each mating. They have to carry unborn children and provide most of the childcare, though females are always certain of the maternity of children. It is also in their interest to get males to invest time and resources during a courtship period, as this increases the chances that the male won't desert and will provide protection and more resources for both the female and her children.

Males on the other hand are never certain of paternity, but invest little in each mating as they are capable of producing millions of sperm and remain fertile over a longer period. Their best strategy is to have as many matings with as many females as possible and to desert at the first opportunity.

Other explanations of aggression include biological ones, which see aggression as being determined solely by physiological factors such as genetics. Moffitt (1992) did a longitudinal study on males aged between 0 and 26 and found that those who had been abused as children and had a low-level activity version of the MAOA gene were more likely to be aggressive. As those who had been abused, but had a high-activity level version of the gene weren't generally aggressive, this suggests the MAOA gene is responsible for aggressive behaviour.

This was backed up by Brunner (1993) who studied the males from one family who had a mutant form of MAOA. They were all retarded and got aggressive when angry, scared or frustrated, suggesting a link between the gene and aggression.

Support comes from Cases (1995) who researched on genetically modified mice lacking the MAOA gene. Their serotonin metabolism was severely affected, leading to increased aggression, even during mating, suggesting a link between genes and aggressive behaviour, though we should be careful when generalising from animals to humans.

🖉 This is a poor answer. In part (a) the outlining of an explanation of group displays is done well, using the example of sport to make pertinent points regarding dominance, motivation, ritualism and intimidation. It is about the right length for the time available, although students are able to write differing amounts.

The answer to part (b) begins well, with the candidate explaining what jealousy is and outlining its role in male–male dominance rivalries. However, after this the candidate does something seen all too often by examiners: they wander off the question. First comes a description of female and male strategies in regard to sexual selection, which is followed by an outline of the contribution that the biological factor of genetics plays in determining aggression. This is nicely evaluated in terms of support from research, with even a considered comment about how representative animal studies are. However, although all perfectly accurate, clearly written and well structured, it is not relevant to the question and as such attracts no credit (infidelity also is not addressed at all). Sometimes this happens because candidates have prepared answers to other questions and offload the material where they can. It is always a good idea to look back

at the question periodically and ask yourself if you are truly answering it in a relevant fashion. As it is, the first sentence accrues some evaluative credit and that is all. However, other explanations of aggression, such as genetics, could be made relevant if clearly used as an evaluative comparison.

(AO1 = (a) 4/5 + (b) AO1 = 2/4) + (AO2/AO3 = 1/16) = 7/25 marks

Development of thinking

(a) Describe Piaget's theory of cognitive development. (9 marks)

(b) Evaluate the applications of Piaget's theory to education. (16 marks)

The descriptive and evaluative parts of this question are divided into two clear parts. However, it is worth pointing out that although part (a) requires a general description of Piaget's theory of cognitive development, part (b) has a narrower focus, requiring the theory to be evaluated in terms of one specific area: its applications to education. This means that general evaluation of the theory will not gain credit, unless such evaluation is made explicitly relevant to applications of the theory to education. The applications of other theories to education, such as Vygotsky's and Bruner's, could also be used as an effective form of comparison.

■ ■ ■

Candidate's answer

(a) Piaget saw cognitive development occurring as a set sequence of stages dependent on biological maturation, with knowledge being actively discovered via interaction with the environment. A neonate has innate schemas, ways of understanding the world, which build up as knowledge is discovered. Any experience fitting an existing schema is assimilated, but those that don't, create disequilibrium; an unpleasant state of imbalance, motivating us to return to equilibrium. This is achieved by accommodating new experiences by altering existing schema. Therefore cognitive development involves swinging between equilibrium and disequilibrium as new experiences are assimilated and accommodated.

The sensorimotor stage occurs between 0 and 2 years, with a child 'thinking' only when it acts on an object. The child is egocentric; making no distinction between itself and the world and object permanence. The realisation that something can occur without being witnessed develops around 18 months of age.

The pre-operational stage occurs from 2 to 7 years with the development of internal images, symbols and language. Children find it hard to classify things in a logical fashion, struggling to understand the relationship between the whole of something and its parts.

In the concrete operational stage from 7 to 11 years, egocentrism declines and a child learns to conserve in a set order (decalage), realising that something can remain the same even if it changes appearance.

The formal operations stage is from 11 to 16 years where abstract reasoning develops and we can think about possibilities and not just actualities.

(b) Piaget's theory was the starting point for a lot of interest and research in cognitive development that has benefited our understanding of how children develop thinking.

Piaget (1954) provided evidence supporting his idea of children lacking object permanence in the sensorimotor stage, when he found that 3- to 4-month-olds wouldn't look for disappeared items. However, Bower and Wishart (1972) think Piaget observed immature motor skills, as 1-month-old babies show surprise when items disappear.

Piaget (1956) provided evidence to show that 7-year-olds were egocentric, because they chose the view they could see of a model mountain when asked to pick a photo of what a doll could see. However, Donaldson thinks this was because they weren't familiar with the landscape and showed they could see from another's viewpoint when using a scenario based on the familiar game of hide and seek.

Support for Piaget comes from the fact that cross-cultural evidence suggests the stages of development are invariant and universal, though formal operations aren't found in all cultures. Indeed the whole concept of separate stages may be too simplistic and misleading, because children can often straddle two stages at once, and Piaget later referred to it as a spiral of development.

A criticism is that Piaget's poor methodology led to him underestimating what children could do and he also overemphasised cognitive aspects at the expense of emotional and social factors. Also Piaget thought development was dependent on biological factors and so couldn't be speeded up, but Meadows (1988) showed that direct tuition could speed up development.

An important difference between Piaget and Bruner concerns the role of language; Piaget sees language ability as dependent on cognitive development, while Bruner sees language development as preceding cognitive development and there is evidence to support both views.

✎ This is a moderate answer. Part (a) provides a concise, clear and accurate outline of Piaget's theory, covering the idea of functional invariants and variant structures, as well as describing the stages of development.

The answer to part (b) is unfortunately all too common and occurs either because candidates misread the question, or because they have only prepared what we get here — the 'general' Piaget essay. The material is evaluative as required, accurate and reasonably detailed. However, the question calls for an assessment of the theory's worth to education and the candidate has failed to do this. Slight credit is awarded for the initial comment, which has some intrinsic value, and later on for the comment about tuition concerning Meadows' research. However, the specification clearly includes application to education and if you are to be certain of being able to answer a question, you must cover everything for that topic listed in the specification.

(AO1 = 9/9) + (AO2/AO3 = 3/16) = 12/25 marks

11

Development of moral understanding

Outline and evaluate one theory of moral understanding. (25 marks)

Kohlberg's theory would probably form the basis of most answers to this question, and it would be a good choice as there is a wealth of material available for both outline and evaluation. For instance, Kohlberg's moral dilemmas and classification system could be used to compose an outline of the theory, while research evidence might be usefully employed to assess its level of support. The contradictory nature of a lot of this evidence might be used to construct a balanced evaluation. As the theory may be affected by age, gender and culture biases, these factors could also be incorporated to build an effective commentary.

■ ■ ■

Candidate's answer

Kohlberg sees morality as a process of biological maturation developing in set stages like cognitive development, though interaction with the environment is important too. Indeed, because he perceives women as restricted to home life, he believes they are less morally developed than men, for which he can be accused of gender bias, and his methodology can be criticised for only initially using male participants. Gilligan widens out the accusation of gender bias to argue that Kohlberg sees morality as based on principles of justice, not applicable to women as they operate on principles of care.

Kohlberg's theory of separate stages involving different types of moral reasoning came about from his research using moral dilemmas. He was not interested in moral opinions, but the thinking behind them, and 72 boys aged between 10 and 16 formed his sample, which was then followed up for years in a longitudinal study. Each boy was questioned about ten moral dilemmas for 2 hours, allowing Kohlberg to decide which stage they were in.

In stage 1, morality is determined by the outcomes of behaviour, not its intentions. In stage 2, moral rules will be followed if it benefits the individual. By stage 3, morality is based on securing the trust and loyalty of others, while stage 4 sees morality as what is best for society. Stage 5 perceives maintaining social order as important, but that bad rules can be changed. People in stage 6 adhere to their own set of moral principles.

Colby (1983) performed follow-up studies for 26 years, finding at age 10 the majority in stage 2, at age 22 the majority in stages 3 and 4 and by age 36 65% were in stage 4, but only 5% had progressed to stage 5. Indeed, as he himself found no evidence of stage 6 in normal participants and less incidence of stage 5 than envisaged, Kohlberg decided that stage 6 might not actually exist. Atkinson (1990) agreed, pointing out that as only 12% of adults reach the post-conventional stages of 5 and 6, it is more of a philosophical ideal than some normal developmental sequence. However, Colby (1983) thought the methodological tool of dilemmas was flawed, as it is impossible when using them to differentiate between stages 5 and 6, so stage 6 might be part of normal development after all.

Kohlberg (1969) tested participants in different cultures, such as Mexico and Taiwan, and found the same invariant pattern of moral development, suggesting it is an innate biological process.

Kohlberg (1975) gave students opportunities to cheat on a test and found that only 15% with higher levels of morality cheated, while 70% of those with lower levels did, suggesting that the level of moral thinking predicts moral behaviour. Fodor (1972) supported this, finding that non-delinquents had a higher level of moral reasoning than delinquents. However, May and Hartshorne (1928) found that when participants were given opportunities to cheat, lie and steal in different situations, moral behaviour was situation specific and not universal across all situations, suggesting that moral reasoning does not reflect behaviour.

Indeed, moral dilemmas are not real life and therefore may be an artificial measuring tool. Gilligan (1982) questioned women facing actual choices of whether to have an abortion and found completely different patterns of moral reasoning than those suggested by Kohlberg, though this may be because she used females.

Although Kohlberg's theory is not without criticism, it has been supported by a lot of research evidence.

✐ This is an excellent answer. A clear, accurate and well detailed outline of the theory is evident, showing how Kohlberg's research led to the theory's formation. A wealth of research evidence is also provided, again with a good level of accurate detail. Evaluation centres on the degree to which evidence supports the theory, with a good balance of positive and negative points being made. Arguments are built up to construct coherent and effective commentaries, with a good awareness of the impact of the methodology used. Although it is not the only way of constructing such a high-level answer — for instance, use could have been made of practical applications and comparisons with other theories — this type of essay is relatively easy to construct, if practised on a regular basis.

(AO1 = 9/9) + (AO2/AO3 = 16/16) = 25/25 marks

Development of social cognition

Outline and evaluate research relating to the Theory of Mind. (25 marks)

The demands of this question are quite clear: outline research relating to the Theory of Mind, for which 9 marks are available, then provide an evaluation, for which a further 16 marks are available. Remember, the term 'research' relates not only to research studies, but also to theoretical aspects. Therefore AO1 credit could be gained by outlining either the theory itself or research studies related to the theory.

The use of the theory as a possible means of explaining autism might form part of an effective evaluation, as well as the theory's ability to demonstrate how children develop the ability to deceive. Research studies themselves could be evaluated in terms of their theoretical support and relevant methodological considerations. As the research in this topic area tends to concern children, there are specific ethical issues that could be explored.

■ ■ ■

Candidate's answer

Animals with a Theory of Mind (ToM) have an awareness of other individuals' mental states, i.e. their thoughts and feelings. It is found in few animal species and humans and is thought to be necessary for social intelligence.

Baron-Cohen and Frith (1985) devised a test for assessing whether an individual had ToM, though it was more to see if autistics had ToM. The test is called the Sally-Anne test and there are two dolls, Sally and Anne, who are watched by a participant. When it is certain the participant knows the dolls' names, Sally puts a marble in her basket and leaves the room. While she is out Anne secretly moves the marble into her box and then Sally returns. The participant is asked where will Sally look for the marble. If they say in Sally's basket they have ToM.

Autistics cannot do this, which explains why they have trouble with social interactions and identifying with characters when reading books.

A criticism of the test is that it involves dolls not real people, but Leslie and Frith (1988) repeated the research using real actors and got the same results, suggesting that the original Sally-Anne test is a valid measure of having ToM.

Interestingly, Down's syndrome children pass the test, suggesting that autism isn't to do with mental retardation and making it even more likely that it is due to a lack of empathy with others.

ToM seems to develop around the age of 4 years all of a sudden, but there is another view that it develops more slowly and gradually. An earlier version of it is called Shared Attention Mechanism (SAM), which appears from 18 months onwards.

Once you've got ToM you can deceive and manipulate people, but you will only develop it when you realise other people can have false beliefs like in the Sally-Anne test. Avis and Harris (1991) found that children of 4 years of age from lots of different cultures in both industrialised countries and non-developed countries could realise people have false beliefs, suggesting ToM is biological and matures when it is ready. It is difficult, though, to conduct cross-cultural research exactly the same with different samples in different cultures and there is a danger of interpreting research in terms of your own cultural viewpoints.

Frith and Frith (1999) used brain scanning to reveal that certain brain areas, like the amygdala and frontal cortex, were activated when individuals were considering other people's mental states. This suggests ToM does have a biological basis and the modular view of ToM is correct.

Flavell (1986) found that 3-year-olds who were given a sponge to handle that looked like a rock, called it a rock, while 4-year-olds called it a sponge.

Bartsch and Wellman (1995) found that ToM seems to appear in both American and Chinese children at the same age of 4 years.

📝 Though slightly on the short side for the time available, this has the makings of a good answer. What is clear is that this candidate understands the requirements of the question and has knowledge of the topic area. However, although theoretical explanations of ToM are given, alongside generally accurate details of research, points are not generally made clear or described in sufficient detail. For example, it is not revealed exactly what the Shared Attention Mechanism actually is, and so there is no indication that the candidate themselves knows. And by the end the candidate appears either rushed or desperately trying to remember anything that may be relevant.

The answer also lacks coherence and at times jumps from one unconnected point to another. Evaluative points are not always made clear: for example, Frith and Frith's research does lend support to the modular theory, but this is in no way explained. Overall this answer would have benefited from a little pre-planning; again this is a skill developed with practice.

(AO1 = 6/9) + (AO2/AO3 = 8/16) = 14/25 marks